RICS NEW RULES OF MEASUREMENT

NRM 2: Detailed measurement for building works

1st edition

Acknowledgments

The project stages from the *RIBA Outline Plan of Work 2007* are reproduced with the permission of the Royal Institute of British Architects.

Published by the Royal Institution of Chartered Surveyors (RICS)

Surveyor Court

Westwood Business Park

Coventry CV4 8JE

UK

www.ricsbooks.com

Produced by the Quantity Surveying and Construction Professional Group of the Royal Institution of Chartered Surveyors.

ISBN 978 1 84219 716 5

Typeset in Great Britain by Columns Design XML Ltd, Reading, Berks

Printed by Page Bros, Norwich

Contents

Foreword

The measurement initiative steering group was set up by the RICS quantity surveying and construction professional group to research the problems associated with the measurement of building works at all stages of the design and construction process. The steering group quickly came to the conclusion that significant improvements where required.

The development units on measurement set up by the measurement initiative steering group have produced a suite of documents covering all aspects of the measurement and description of a building project – from 'cradle to grave'.

This volume, *RICS new rules of measurement: Detailed measurement for building works* ('NRM 2'), provides fundamental guidance on the quantification and description of building works for the purpose of preparing bill of quantities and quantified schedules of works; it also provides a sound basis for designing and developing standard or bespoke schedules of rates. Direction on how to deal with items that are unquantifiable is also provided – e.g. preliminaries, overheads and profit, contractor designed works, risk transfer and fluctuations.

The *RICS new rules of measurement: Detailed measurement for building works* provides a uniform basis for measuring and describing building works and embodies the essentials of good practice.

NRM 2 replaces the *Standard Method of Measurement for Building Works* ('SMM'), which was first published by RICS in 1922, with the latest edition being SMM7, published in 1988.

This first edition will become operative on 1 January 2013 and is post dated accordingly.

Acknowledgments

The development of the *RICS new rules of measurement: Detailed measurement for building works* was facilitated by the RICS Quantity Surveying and Construction Professional Group under the direction of the steering group.

Lead authors:

- David Benge, Gleeds Cost Management Limited
- John Davidson, Consultant

Contributors:

- Simon Cash, Appleyards
- Graham Hadden, V B Johnson
- Michael Jewell, MDA Consulting
- Michael Rickard (Harvey Frost)
- Justin Sullivan, Adair Associates
- Keith Tweedy

Steering Group:

- **Chair**: Stuart Earl, Gleeds Cost Management Limited
- David Benge, Gleeds Cost Management Limited
- Alan Cripps, RICS
- John Davidson, Consultant
- Adrian Green, Laing O'Rourke
- Andy Green, Faithful and Gould
- Joe Martin, BCIS
- Alan Muse, RICS
- Sean Smylie, Shepherd Construction; Contractors Legal Group; UK Contractors Group
- Roy Stratton, Countryside Properties

The Steering Group express their thanks to the professional institutions and trade bodies, and to the quantity surveyors and building contractors for their co-operation and advice in the detailed consultations that have taken place. The steering group and development unit also express their thanks to the surveyors who have assisted in testing the *RICS new rules of measurement: Detailed measurement for building works*, and to the editors who have had the task of bringing together the final document.

Introduction

Status of the *RICS new rules of measurement*

These rules have the status of a guidance note. Where recommendations are made for specific professional tasks, these are intended to represent 'best practice', i.e. recommendations which in the opinion of RICS meet a high standard of professional competence.

Although members are not required to follow the recommendations contained in the note, they should take into account the following points.

When an allegation of professional negligence is made against a surveyor, a court or tribunal may take account of the contents of any relevant guidance notes published by RICS in deciding whether or not the member had acted with reasonable competence.

In the opinion of RICS, a member conforming to the practices recommended in this note should have at least a partial defence to an allegation of negligence if they have followed those practices. However, members have the responsibility of deciding when it is inappropriate to follow the guidance.

It is for each surveyor to decide on the appropriate procedure to follow in any professional task. However, where members do not comply with the practice recommended in this note, they should do so only for a good reason. In the event of a legal dispute, a court or tribunal may require them to explain why they decided not to adopt the recommended practice. Also, if members have not followed this guidance, and their actions are questioned in an RICS disciplinary case, they will be asked to explain the actions they did take and this may be taken into account by the Panel.

In addition, guidance notes are relevant to professional competence in that each member should be up to date and should have knowledge of guidance notes within a reasonable time of their coming into effect.

Document status defined

RICS produces a range of standards products. These have been defined in the table below. This document is a guidance note.

Type of document	Definition	Status
RICS practice statement	Document that provides members with mandatory requirements under Rule 4 of the Rules of Conduct for members	Mandatory
RICS code of practice	Standard approved by RICS, and endorsed by another professional body that provides users with recommendations for accepted good practice as followed by conscientious practitioners	Mandatory or recommended good practice (will be confirmed in the document itself)
RICS guidance note	Document that provides users with recommendations for accepted good practice as followed by competent and conscientious practitioners	Recommended good practice
RICS information paper	Practice based information that provides users with the latest information and/or research	Information and/or explanatory commentary

The RICS new rules of measurement (NRM) suite of documents in context

The *RICS new rules of measurement* (NRM) is a suite of documents issued by the RICS Quantity Surveying and Construction Professional Group. The rules have been written to provide a standard set of measurement rules that are understandable by anyone involved in a construction project.

The rules provide essential guidance to all those involved in, as well as those who wish to be better informed about, the cost management of construction projects. Although the *RICS new rules of measurement* are principally based on UK practice, the requirements for a coordinated set of rules and underlying philosophy behind each volume have worldwide application.

- NRM 1: Order of cost estimating and cost planning for capital building works
- NRM 2: Detailed measurement for building works
- NRM 3: Order of cost estimating and cost planning for building maintenance works.

The NRM suite comprises the following three volumes:

NRM 1: Order of cost estimating and cost planning for capital building works

This volume provides indispensable guidance on the quantification of building works for the purpose of preparing cost estimates and cost plans. Direction on how to quantify other items forming part of the cost of a construction project, but which are not reflected in the measurable building work items, is also provided – i.e. preliminaries, overheads and profit, project team and design team fees, risk allowances, inflation, and other development and project costs.

NRM 1 is the 'cornerstone' of good cost management of construction projects – enabling more effective and accurate cost advice to be given to clients and other project team members, as well as facilitating better cost control.

Although written primarily for the preparation of order of cost estimates and cost plans, the rules will be invaluable when preparing approximate estimates.

In addition, the rules can be used as a basis for capturing historical cost data in the form required for order of cost estimates and elemental cost plans, thereby completing the 'cost management cycle'.

NRM 2: Detailed measurement for building works

This volume provides fundamental guidance on the detailed measurement and description of building works for the purpose of obtaining a tender price. The rules address all aspects of bill of quantities (BQ) production, including setting out the information required from the employer and other construction consultants to enable a BQ to be prepared, as well as dealing with the quantification of non-measurable work items, contractor designed works and risks. Guidance is also provided on the content, structure and format of BQ, as well as the benefits and uses of BQ.

While written mainly for the preparation of bill of quantities, quantified schedules of works and quantified work schedules, the rules will be invaluable when designing and developing standard or bespoke schedules of rates.

These rules provide essential guidance to all those involved in the preparation of bill of quantities, as well as those who wish to be better informed about the purpose, use and benefits of bill of quantities.

NRM 3: Order of cost estimating and cost planning for building maintenance works

This volume provides essential guidance on the quantification and description of maintenance works for the purpose of preparing initial order of cost estimates during the preparation stages of a building project, cost plans during the design development and pre-construction stages, and detailed, asset-specific cost plans during the pre-construction phases of a building project. The guidance provided by the rules also aids the procurement and cost control of maintenance works.

The rules follow the same framework and premise as *NRM 1: Order of cost estimating and cost planning for capital building works*. Consequently, they give direction on how to quantify and measure other items associated with maintenance works, but which are not reflected in the measurable maintenance work items – i.e. maintenance contractor's management and administration charges, overheads and profit, other maintenance-related costs, consultants' fees and risks in connection with maintenance works.

Unlike capital building works projects, maintenance works are required to be carried out from the day a building or asset is put to use until the end of its life. Accordingly, while the costs of a capital building works project are usually incurred by the building owner/developer over a relatively short-term, costs in connection with maintenance works are incurred throughout the life of the building – over the long-term. Consequently, the rules provide guidance on the measurement and calculation of the time value of money, guidance on using the measured data to inform life cycle of cost plans and forward maintenance plans, as well as VAT, and taxation.

NRM 3: Order of cost estimating and cost planning for building maintenance works, together with *NRM 1: Order of cost estimating and cost planning for capital building works*, presents the basis of good cost management of capital building works and maintenance works – enabling more effective and accurate cost advice to be given to clients and other project team members, as well as facilitating better cost control.

Identity

The rules are called *NRM 2: Detailed measurement for building works*.

Enquiries

Enquiries concerning the rules may be made in the first instance to the Quantity Surveying and Construction Professional Group at RICS.

RICS QS & Construction Professional Group

The Royal Institution of Chartered Surveyors (RICS)

Parliament Square

London SW1P 3AD

United Kingdom

e: qsandc.professionalgroup@rics.org

Any suggestions for future revisions are welcomed and may also be sent to the Quantity Surveying and Construction Professional Group at RICS.

Part 1: General

Part 1: General

1.1 Introduction

1.1.1 This part places detailed measurement for bill of quantities, schedules of works and work schedules in context with the RIBA Outline Plan of Work and the OGC Gateway Process; and explains the symbols, abbreviations and definitions used in the rules.

1.2 Measurement in context with the RIBA Plan of Work and OGC Gateway Process

1.2.1 Throughout this document, references are made to both the RIBA Plan of Work and the OGC Gateway Process and the RIBA Work Stages/OGC Gateways within.

1.2.2 The RIBA Plan of Work is a construction industry recognised framework that organises the process of managing and designing building projects and administering building contracts into a number of key work stages. The RIBA Plan of Work has 11 sequential steps known as 'RIBA Work Stages'. Despite its apparent linear nature, it should be recognised that the sequence or content of the RIBA Work Stages may need to be varied or overlapped to suit the proposed procurement method. Consequently, when two or more RIBA Work Stages are combined, it is not always transparent when a building project is moving from one stage to another. As such, it is an ideal tool, provided that it is conceptualised as providing the basic outline of the building project process.

1.2.3 As an alternative to the RIBA Plan of Work, central civil government, the health sector, local government and the defence sector have adopted the OGC Gateway Process as best practice for managing and designing building projects. The process examines programmes and projects at key decision-points in their lifecycle. It looks ahead to provide assurance that the employer can progress to the next stage. Project reviews are carried out under OGC Gateway reviews 1 to 5. Typically, a project will undergo three reviews before commitment to invest, and two looking at service implementation and confirmation of the operational benefits.

1.2.4 Both the RIBA Plan of Work and the OGC Gateway Process are recognised frameworks for managing and designing building projects.

1.2.5 The point at which measurement is carried out by the quantity surveyor/cost manager for the purpose of preparing bill of quantities, or quantified schedule of works, in the context of the RIBA Work Stages and OGC Gateways is shown in Figure 1.1.

1.2.6 Bill of quantities are prepared by the quantity surveyor/cost manager at RIBA Work Stage G (Tender Documentation).

1.2.7 The project stages from the *RIBA Outline Plan of Work 2007* (overleaf) are reproduced with the permission of the Royal Institute of British Architects.

Figure 1.1: The RICS formal cost estimating and cost planning stages in context with the RIBA Plan of Work and OGC Gateways (adapted from the RIBA Outline Plan of Work 2007).

RIBA Work Stages			RICS cost estimating, elemental cost planning and tender document preparation stages	OGC Gateways (Applicable to projects)	
Preparation	A	Appraisal	Order of cost estimates (as required to set authorised budget)	1	Business Justification
Preparation	B	Design Brief		2	Delivery Strategy
Design	C	Concept	Formal cost plan 1	3A	Design Brief and Concept Approval
Design	D	Design Development	Formal cost plan 2		
Design	E	Technical Design		3B	Detailed Design Approval
Pre-construction	F	Production Information	Formal cost plan 3 Pre-tender estimate		
Pre-construction	G	Tender Documentation	Bills of quantities (Quantified) schedule of works (Quantified) work schedules		
Pre-construction	H	Tender Action	Post tender estimate	3C	Investment Decision
Construction	J	Mobilisation			
Construction	K	Construction to Practical Completion		4	Readiness for Service
Use	L	Post Practical Completion		5	Operational Review and Benefits Realisation

Note: A prerequisite of OGC Gateway Review 3: Investment Decision, is that the design brief, concept design and detailed design have been approved and signed off by the Senior Responsible Owner (SRO). For the purpose of comparing the OGC Gateway Process with the RIBA Work Stages, these two decision points are referred to as OGC Gateway 3A (Design Brief and Concept Approval) and OGC Gateway 3B (Detailed Design Approval); with OGC Gateway 3C representing the final OGC Gateway Review 3 (Investment Decision).

1.3 Purpose of NRM 2

1.3.1 *RICS new rules of measurement: Detailed measurement for building works* has been written to provide a standard set of measurement rules for the procurement of building works that are understandable by all those involved in a construction project, including the employer; thereby aiding communication between the project/design team and the employer.

1.3.2 The document provides rules of measurement for the preparation of bill of quantities and schedules of works (quantified). The rules also provide a framework, which can be used to develop bespoke and standard schedules of rates.

1.3.3 The rules address all aspects of bill of quantities (BQ) production, including setting out the information required from the employer and other construction consultants to enable a BQ to be prepared, as well as dealing with the quantification of non-measurable work items, contractor designed works and risks. Guidance is also provided on the content, structure and format of BQ

1.3.4 The *RICS new rules of measurement* suite of documents is based on UK practice but the requirements for a coordinated set of rules and underlying philosophy behind each section have worldwide application.

1.4 Use of NRM 2

1.4.1 The *RICS new rules of measurement: Detailed measurement for building works* provides a structured basis for measuring building work and presents a consistent approach for dealing with other key cost components associated with a building project when preparing bill of quantities and quantified schedule of works. The rules represent the essentials of good practice.

1.4.2 The rules address both the production of bill of quantities (BQ) for entire building project and for discrete work packages. They can also be used to prepare quantified schedules of works and quantified work schedules.

1.4.3 In addition, the framework can be used to construct both bespoke and standard schedules of rates for the purpose of:

(1)   Discrete contracts;

(2)   Term contracts; and

(3)   Framework arrangements.

1.4.4 Users of the rules are advised to adopt metric units as the standard system of measurement. Where the employer requires reference to imperial units these may be provided as supplementary information (e.g. in parenthesis).

1.4.5 Although the British Standard BS 8888:2006 *Technical Product Specification* (for defining, specifying and graphically representing products) recommends the inclusion of a comma rather than a point as a decimal marker, and a space instead of a comma as a thousands separator, the traditional UK convention has been adopted in these rules (i.e. a point as a decimal marker and a comma as a thousands separator). Users should take care to ensure that this does not conflict with employer requirements.

1.5 Structure of NRM 2

1.5.1 This document is divided into three parts with supporting appendices:

- **Part 1** places the places measurement for works procurement in context with the *RIBA Plan of Work* and the *OGC Gateway Process*; and explains the symbols, abbreviations and definitions used in the rules.

- **Part 2** outlines the benefits of detailed measurement, describes the purpose and uses of *RICS new rules of measurement: Detailed measurement for building works*; explains the function of bill of quantities, provides work breakdown structures for bill of quantities, defines the information required to enable preparation of bill of quantities, describes the key constituents of bill of quantities, explains how to prepare bill of quantities; sets out the rules of measurement for the preparation of bill of quantities; and provides the method for dealing with contractors' preliminaries, contractors' overheads and profit, contractors' design fees, other development/project costs, risks, inflation, and data-gathering for supporting claims for tax incentives.

- **Part 3** comprises the tabulated rules for the measurement and description of building works for the purpose of works procurement.

- **Appendices**:

  Appendix A: Guidance on the preparation of bill of quantities

  Appendix B: Template for preliminaries (main contract) pricing schedule (condensed)

  Appendix C: Template for preliminaries (main contract) pricing schedule (expanded)

  Appendix D: Template for pricing summary for elemental bill of quantities (condensed)

  Appendix E: Template for pricing summary for elemental bill of quantities (expanded)

  Appendix F: Templates for provisional sums, risks and credits

  Appendix G: Example of a work package breakdown structure

1.6 Symbols, abbreviations and definitions

Symbols, abbreviations and certain key words and phrases used in the rules are detailed below.

1.6.1 Symbols used for measurement

ha	hectare
hr	hour
kg	kilogramme
kN	kilonewton
kW	Kilowatt
m	linear metre
m^2	square metre
m^3	cubic metre
mm	millimetre
mm^2	square millimetre
mm^3	cubic millimetre
nr	number
t	tonne
wk	week

1.6.2   Abbreviations

BQ	bill of quantities
BQBS	bill of quantities (or BQ) breakdown structure
CBS	cost breakdown structure
OGC	Office of Government Commerce
PC sum	prime cost sum
PC price	prime cost price
RIBA	Royal Institute of British Architects
RICS	Royal Institution of Chartered Surveyors
WBS	work breakdown structure

Note: The names of UK government departments are frequently changed. This applies to the Office of Government Commerce. Notwithstanding this, the acronym OGC has not been changed as it is a construction industry recognised phrase.

1.6.3   Definitions

Bill of quantities (BQ) – means a list of items giving detailed identifying descriptions and firm quantities of the work comprised in a contract.

Cost breakdown structure (CBS) – in the context of bill of quantities, represents the financial breakdown of a building project into cost targets for elements or work packages.

Cost target – in the context of bill of quantities, means the total expenditure for an element or work package.

Credit – means a refund offered by the contractor to the employer in return for the benefit of taking ownership of materials, goods, items, mechanical and electrical plant and equipment, etc. arising from demolition or strip out works.

Daywork – means the method of valuing work on the basis of time spent by the contractor's workpeople, the materials used and the plant employed.

Defined provisional sum – means a sum provided for work which is not completely designed but for which the following information shall be provided:

–   the nature and construction of the work;

–   a statement of how and where the work is fixed to the building and what other work is to be fixed thereto;

–   a quantity or quantities that indicate the scope and extent of the work; and

–   any specific limitations and the like identified.

Design team – means architects, engineers and technology specialists responsible for the conceptual design aspects of a building, structure or facility and their development into drawings, specifications and instructions required for construction and associated processes. The design team is a part of the project team.

Director's adjustment – means a reduction or addition to the tender price, derived by the contractor's estimating team, offered by the director(s) of the contractor.

Employer – means the owner and/or the developer of the building; in some cases the ultimate user. The terms Senior Responsible Owner (SRO) and project sponsor are used by central civil government and the defence sector; being the representatives empowered to manage the building project and make project-specific decisions. For the purpose of these measurement rules, the term 'employer' will also be used interchangeably with Senior Responsible Owner (SRO) or project sponsor.

Fixed charge – is for work, the cost of which is to be considered independent of duration.

Main contractor (or **prime contractor**) – means the contractor responsible for the total construction and completion process of the building project. The term 'prime contractor' is often used in central civil government and the defence sector to mean 'main contractor'.

Main contract preliminaries – are items that cannot be allocated to a specific element, sub-element or component. Main contract preliminaries include the main contractor's costs associated with management and staff, site establishment, temporary services, security, safety and environmental protection, control and protection, common user mechanical plant, common user temporary works, the maintenance of site records, completion and post-completion requirements, cleaning, fees and charges, sites services and insurances, bonds, guarantees and warranties. Main contractors' preliminaries exclude costs associated with subcontractors' or work package contractors' preliminaries.

OGC Gateway Process – is a process that examines programmes and projects at key decision-points in their lifecycle. It looks ahead to provide assurance that the employer can progress to the next stage. Project reviews are carried under OGC Gateway reviews 1 to 5. Typically, a project will undergo three reviews before a commitment to invest, and two looking at service implementation and confirmation of the operational benefits. The process is best practice in central civil government, the health sector, local government and the defence sector. The emphasis of the OGC Gateway Process is to examine the business case, which requires an assessment of the total development cost of the building project.

OGC Gateways (or OGC Gateway) – are key decision points within the OGC Gateway Process.

Other development/project costs – means costs that are not necessarily directly associated with the cost of constructing the building, but form part of the total cost of the building project to the employer (e.g. land acquisition costs, fees for letting agents, marketing costs and contributions associated with planning permissions).

Overheads and profit – means the contractor's costs associated with head office administration, proportioned to each building contact, plus the main contractor's return on capital investment.

Post tender estimate – means a cost estimate carried out after the evaluation of tenders to corroborate the funds required by the employer to complete the building project.

Pre-tender estimate – means a cost estimate prepared immediately before calling tenders for construction.

Preliminaries – see main contract preliminaries and work package contract preliminaries.

Provisional quantity – means a quantity which cannot be accurately determined (i.e. an estimate of the quantity).

Provisional sum – means a sum of money set aside to carry out work that cannot be described and given in quantified items in accordance with the tabulated rules of measurement. A provisional sum will be identified as either 'defined' or 'undefined' (see definitions of 'Defined provisional sums' and 'Undefined').

Prime cost sum (PC Sum) – means a sum of money included in a unit rate to be expended on materials or goods from suppliers (e.g. ceramic wall tiles at £36.00/m or door furniture at £75.00/door). It is a supply only rate for materials or goods where the precise quality of those materials and goods are unknown. PC Sums exclude all costs associated with fixing or installation, all ancillary and sundry materials and goods required for the fixing or installation of the materials or goods, subcontractor's design fees, subcontractor's preliminaries, subcontractor's overheads and profit, Main Contractor's design fees, preliminaries and overheads and profit.

Residual risk (or **retained risk**) – means the risks retained by the employer.

RIBA Outline Plan of Work summarises the deliverables required under each RIBA Work Stage.

RIBA Plan of Work – is a model procedure dealing with basic steps in decision making for a medium-sized building project. The RIBA Plan of Work sets out a logical structure for building projects, starting with the brief and ending with post-occupancy evaluation. The procedures identify the responsibilities of the design team at each stage of the design and contract administration process. Each step is referred to as a RIBA Work Stage. The full title of the RIBA Plan of Work is *The Architect's Plan of Work*, published by RIBA, but it is commonly known and referred to as the RIBA Plan of Work in the building construction industry.

RIBA Work Stage – means the stage into which the process of designing building projects and administering building contracts may be divided. Some variations of the RIBA Work Stages apply for design and build procurement.

Statutory undertaker – means organisations, such as water, gas, electricity and telecommunications companies, that are authorised by statute to construct and operate public utility undertakings.

Subcontractor – means a contractor employed by the main contractor to undertake specific work within the building project; also known as specialist, works, trade, work package, and labour only contractors.

Time-related charge – is for work, the cost of which is to be considered dependent on duration.

Undefined provisional sum – means a sum provided for work that is not completely designed, but for which the information required for a defined provisional sum cannot be provided.

Work breakdown structure (WBS) – in the context of bill of quantities, is used to sub-divide a building project into meaningful elements or work packages.

Work package contractor – a specialist contractor who undertakes particular identifiable aspects of work within the building project; e.g. ground works, cladding, mechanical engineering services, electrical engineering services, lifts, soft landscape works or labour only. Depending on the contract strategy, works contractors can be employed directly by the employer or by the main contractor.

Work package contract preliminaries – are preliminaries that relate specifically to the work that is to be carried out by a work package contractor.

Part 2: Rules for detailed measurement of building works

Part 2: Rules for detailed measurement of building works

2.1 Introduction

2.1.1 Part 2 of the rules describes the purpose and uses of *NRM 2: Detailed measurement for building works*; describes the types of bill of quantities (BQ); gives guidance on the preparation and composition of BQ; and defines the information required to enable a BQ to be prepared. Part 2 also sets out the rules of measurement of building items, and the rules for dealing with preliminaries, non-measurable works, and contractor designed works, as well as risks, overheads and profit, and credits.

2.1.2 In addition, Part 2 of the rules deals with other aspects of BQ production, including price fluctuations, director's adjustments, daywork and value added tax. Guidance is also provided on the codification of BQ, the use of BQ for cost control and cost management, and the analysis of a BQ to provide cost data.

2.2 Purpose of bill of quantities

2.2.1 The primary purposes of a bill of quantities (BQ), which becomes a contract document, are:

- to provide a co-ordinated list of items, together with their identifying descriptions and quantities, that comprise the works to enable contractors to prepare tenders efficiently and accurately; and
- when a contract has been entered into, to:
  - provide a basis for the valuation of work executed for the purpose of making interim payments to the contractor; and
  - provide a basis for the valuation of varied work.

2.2.2 Essentially, a BQ is a list of the items with detailed identifying descriptions and quantities, which make up the component parts of a building.

2.3 Benefits of bill of quantities

2.3.1 Irrespective of what contract strategy is used, at some stage in the procurement process one party will need to quantify the extent of works to be executed; whether it be the employer's quantity surveyor/cost manager, the main contractor or the work package contractors for the purpose of obtaining a price for completing building works, valuing the extent of work complete for purposes of payment, valuing variations in the content or extent of building works, or to support applications for tax or other financial incentives. Consequently, detailed measurement for the purpose of bill of quantities (BQ) production is beneficial for a number of reasons:

- it saves the cost and time of several contractors measuring the same design in order to calculate their bids for competition;
- it provides a consistent basis for obtaining competitive bids;
- it provides an extensive and clear statement of the work to be executed;
- it provides a very strong basis for budgetary control and accurate cost reporting of the contract (i.e. post contract cost control), including:
  - the preparation of cash flow forecasts,
  - a basis for valuing variations, and
  - a basis for the preparation of progress payments (i.e. interim payments);
- it allows, when BQ items are codified, reconciliation and any necessary transfers and adjustments to be made to the cost plan;

- when priced, it provides data to support claims for tax benefits (e.g. capital allowances and value added tax (VAT));
- when priced, it provides data to support claims for grants; and
- it provides one of the best sources of real-time cost data, which can be used for estimating the cost of future building projects (i.e. historic cost information (see 2.17)) as it provides a cost model in a single document.

2.4 Types of bill of quantities

2.4.1 The use of bill of quantities in support of a contract is the traditional and proven means of securing a lump-sum price for undertaking building works. Bill of quantities (BQ) can be:

- firm (to obtain a lump-sum price for a fully designed building project); or
- approximate (subject to remeasurement as built).

2.4.2 **Firm bill of quantities:**

2.4.2.1 The reliability of the tender price will increase in relation to the accuracy of the quantities provided (i.e. the more precisely the work is measured and described). In theory, were there no design changes, then a firm BQ would provide a price at tender stage, which would equal the final cost. In practice there will be changes, and the BQ provides a good basis for cost control, since the direct cost of change can be assessed with reference to the BQ rates.

2.4.2.2 The firmer the BQ the better it is as a means of financial control.

2.4.3 **Approximate bill of quantities:**

2.4.3.1 Approximate BQ are used when there is insufficient detail to prepare firm BQ or where it is decided by the employer that the time or cost of a firm BQ is not warranted. Such contracts do not provide a lump-sum price, but rather tender price totals (i.e. a quantified schedule of rates), since the quantities are subject to re-measurement on completion by the quantity surveyor/cost manager. These contracts are usually subject to greater variation than lump sum contracts and therefore should only be used where time is a limiting factor or where there is great uncertainty in respect of certain elements, such as major excavation and earthworks.

2.4.3.2 The initial resource cost of an approximate BQ is likely to be lower than for a firm BQ, but the need for re-measurement invariably results in an overall higher resource cost.

2.4.3.3 Although the quantities are approximate, the descriptions of work items should be correct.

2.5 Preparation of bill of quantities

2.5.1 Bill of quantities (BQ) are produced at 'RIBA Work Stage G (Tender Documentation)' or as an intrinsic part of 'OGC Gateway 3C (Investment Decision)'. The requirements of RIBA Work Stage G, as described in the RIBA Outline Plan of Work, are as follows:

> *Preparation and/or collation of tender documentation in sufficient detail to enable a tender or tenders to be obtained for the project.*

2.5.2 To enable the preparation of BQ, the information resulting from RIBA Work Stages E (Technical Design) and F (Production Information) will be required. The requirements of RIBA Work Stages E and F, as described in the RIBA Outline Plan of Work, are as follows:

> *Preparation of technical design(s) and specifications, sufficient to co-ordinate components and elements of the project and information for statutory standards and construction safety.*
>
> *F1 Preparation of production information in sufficient detail to enable a tender or tenders to be obtained.*

> *F2 Application for statutory approvals. Preparation of further information for construction required under the building contract.*
>
> *Project stages from the RIBA Outline Plan of Work 2007 (Amended November 2008), copyright Royal Institute of British Architects, are reproduced here with the permission of the RIBA.*

2.5.3 The processes defined by RIBA Work Stages E (Technical Design), F (Production Information) and G (Tender Documentation) are commensurate with the processes required to meet the requirements of OGC Gateway 3C (Investment Decision).

2.5.4 The information and documents required for the preparation of BQ are described in paragraph 2.14 of these rules.

2.5.5 BQ required for a lump sum contract based on firm or approximate quantities will normally be prepared by the employer's quantity surveyor/cost manager, whereas under a design and build contract, the employer's project team will prepare the 'employer's requirements' and the BQ or quantified schedules of work will be prepared by either the main contractor or, more likely, the main contractor's work package contractors. The choice of who quantifies building works is solely down to the employer's preference of contract strategy (see Figure 2.1 below).

Figure 2.1: Responsibility for quantifying building works

Contract strategy	Basis of 'Invitation Documents'	Preparation by
Traditional lump sum	(a) Firm BQ	Employer's quantity surveyor/cost manager.
	(b) Approximate BQ	
Design and build	Employer's Requirements	Employer's project team (with compilation normally by the employer's quantity surveyor/cost manager). **Note:** Quantification of the Employer's Requirements will be carried out by either the main contractor or work package contractors; who will prepare firm or approximate BQ, or quantified schedules of work as appropriate.
Management	(a) Firm BQ	Employer's quantity surveyor/cost manager (or main contractor or work package contractors if 'Invitation Documents' prepared by employer's quantity surveyor/cost manager or main contractor, respectively, are based on either 'specification and drawings' or 'un-quantified schedule of works' (i.e. un-quantified information).
	(b) Approximate BQ	
Management (design and manage)	(a) Firm BQ	
	(b) Approximate BQ	
Construction management	(a) Firm BQ	
	(b) Approximate BQ	

2.5.6 Guidance on the preparation of a bill of quantities (BQ) is given in Appendix A of these rules.

2.6 Composition of a bill of quantities

2.6.1 Bill of quantities (BQ) usually comprise the following sections:
- Form of Tender (including certificate of bona fide tender);
- Summary (or Main Summary);
- Preliminaries, comprising two sections as follows:
  - Information and requirements; and
  - Pricing schedule;
- Measured work (incorporating contractor designed works);
- Risks;

- Provisional sums;
- Credits (for materials arising from the works);
- Dayworks (Provisional); and
- Annexes.

2.6.2 **Form of tender:**

2.6.2.1 This is a document that is used to record the main contractor's price for completing the building project (i.e. his tender price). If accepted by the employer, the tender price will become the 'contract sum'. The form of tender can be a separate document.

2.6.2.2 A separate 'certificate of bona fide tender', which is completed by the main contractor to confirm that he has not communicated his tender to other parties, is sometimes inserted after the form of tender. Alternatively, the employer's requirements for confirming that bona fide tender has been submitted by the main contractor can be incorporated in the form of tender.

2.6.3 **Summary (or main summary):**

2.6.3.1 The summary, sometimes called the main summary, is made at either the front or end of the bill of quantities and comprises a list of the bill that form the entire bill of quantities. The total price for each section of the bill of quantities (BQ) is carried forwarded and inserted against the applicable item listed in the summary. For example, a summary for an elemental bill will address all or some of the following:
- Preliminaries;
- Measured works (including 'Contractor Designed Works':
    - Facilitating works
    - Substructure
    - Superstructure
    - Internal finishes
    - Fittings, furnishings and equipment
    - Services
    - Complete buildings
    - Works to existing buildings
    - External works
- Risks;
- Provisional sums:
    - Defined
    - Undefined
- Works to be carried out by statutory undertakers;
- Overheads and profit;
- Credits (for materials arising from the works);
- Fixed price adjustment;
- Director's adjustment;
- Dayworks (Provisional);
- Total price (to Form of Tender).

2.6.3.2 Where the measured work has been divided into work sections, the work sections will be listed instead of elements.

2.6.3.3 At the end of the summary, provision is made to total the list to ascertain the total price and transfer the total to the form of tender, which, subject to verification and any necessary adjustments, will become the contract sum referred to by the conditions of contract.

2.6.4 Preliminaries:

2.6.4.1 Preliminaries address and communicate to the contractor items that are not directly related to any component, element, or work section (i.e. measured works). The information provided will enable the contractor to ascertain his or her price for, among other things, management of the building project, site establishment, security, safety, environmental protection and common user mechanical plant, as well as the employer's completion and post-completion requirements. Preliminaries are divided into two sections as follows:

2.6.4.2 *Information and requirements:*

The purpose of the information and requirements section is to describe the building project particulars; the drawings upon which the bill of quantities was based; the work in general; the site and any adjoining properties; the form of contract and any amendments and/or supplementary conditions to the form of contract; the employer's specific requirements; and any specific limitations or restrictions that might impact on the sequence and/or method of working. It also contains information on how to interpret the bill of quantities, including any special methods of measurement (i.e. where the method of measurement has deviated from *NRM 2: Detailed measurement for building works*).

2.6.4.3 *Pricing schedule:*

The pricing schedule is simply a schedule in which the contractor inserts preliminary costs relating to the employer's requirements, and all contractor cost items (including management and staff, site accommodation, services and facilities, mechanical plant, and temporary works items). The quantification of preliminaries is dealt with in paragraph 2.7.

2.6.5 Measured work:

2.6.5.1 This is the main part of the bill of quantities, which lists all the items of work to be undertaken. The quantities and descriptions of items should be determined in accordance with the tabulated rules of measurement in Part 3 (Tabulated rules for the measurement of building works) of these rules. Various methods can be used to present the measured work (i.e. bill breakdown structures – see paragraph A.1 at Appendix A: Guidance on the preparation of bill of quantities).

2.6.5.2 The rules relating to the quantification and description of measured work are given in paragraph 2.8.

2.6.6 Risks:

2.6.6.1 This section comprises a list of residual risks (i.e. unexpected expenditure arising from risks that materialise, for example, disposal of contaminated ground material), which the employer wishes to transfer to the contractor. The contractor is required to provide a lump-sum fixed price for taking, managing and dealing with the consequences of the identified risk should it materialise.

2.6.6.2 The rules relating to the quantification and description of risk are given in paragraph 2.10.

2.6.7 Provisional sums:

2.6.7.1 Provisional sums are sums included for any items of work that are anticipated, but for which no firm design has been developed, including any sums listed for any items of work that are to be executed by a statutory undertaker.

2.6.7.2 This part of the BQ, therefore, lists items of work that cannot be entirely foreseen or detailed accurately at the time tenders are invited (i.e. non-measurable items). Pre-determined sums of money are set against each item, determined by the quantity surveyor/cost manager, to cover their cost.

2.6.7.3 The rules relating to provisional sums are given in paragraph 2.9.1.

2.6.8 **Credits (for materials arising from the works):**

2.6.8.1 This section of the BQ comprises a list of materials arising from the works for which the employer requires the contractor to offer a credit.

2.6.8.2 The rules relating to the quantification and description of credits are given in paragraph 2.12.

2.6.9 **Dayworks (Provisional):**

2.6.9.1 This section of the BQ gives provision for the contractor to competitively tender rates and prices for works, which might be instructed to be carried out on a daywork basis.

2.6.9.2 The rules relating to time-charged work carried out on the basis of daywork rates are given in paragraph 2.13.3 below.

2.6.10 **Annexes:**

The annexes comprise information referred to in the BQ where it is not contained in, or to be issued as, a separate document. Examples are: performance specifications (if not included in project specification); copies of quotations; and copies of communications with statutory undertakers.

2.7 Preliminaries

2.7.1 For the most part, preliminaries are the cost of administering a project and providing plant, site staff, facilities site-based services, and other items not included in the rates for measured works.

2.7.2 Preliminaries are divided into two categories:
(1) preliminaries (main contract)
(2) preliminaries (works package contract).

2.7.3 **Preliminaries (main contract)** are divided into two discrete sections:
(1) information and requirements; and
(2) pricing schedule.

2.7.3.1 *Information and requirements:*

This is the descriptive part of the preliminaries (main contract), which:
(1) sets out the project particulars (e.g. the project title, the site address and the names and contact details of the employer and the employer's project team);
(2) identifies the drawings upon which the bill of quantities was based;
(3) depicts the boundary of the construction site;
(4) provides information about existing buildings and existing mains services on or adjacent to the site, and about any existing records that will inform the main contractor of any known or potential hazards that need to be considered;
(5) identifies known constraints and restrictions that might impact on the main contractor's methodology for constructing the building or buildings;
(6) describes the building project in general;
(7) specifies the standard form of contract, together with the contract particulars and any amendments and/or supplementary or special conditions to the standard form of contract, which is to be entered into by the contracting parties; as well as the employer's requirements in respect of insurances, parent company guarantees, performance bonds and collateral warranties;
(8) explains the documents provided, their content and how they are to be used;

(9) confirms the method of measurement (i.e. *NRM 2: Detailed measurement for building works*), how to interpret the bill of quantities, and any special methods of measurement (i.e. where the method of measurement has deviated from the specified rules);

(10) sets out the employer's specific requirements in respect of:

    (a) management of the works by the main contractor, including progress reporting, programme management and cost reporting requirements;

    (b) quality standards to be achieved, and quality control requirements, by the main contractor, including inspection, testing and commissioning requirements;

    (c) security, safety and protection measures to be provided by the main contractor;

    (d) facilities, temporary works and services required by the employer;

    (e) specific requirements such as advertising, the provision of a marketing suite or a topping out event;

    (f) specific limitations on method, sequence, timing of the 'works' imposed by the employer, including out of normal hours working and phasing requirements;

    (g) operation and maintenance of the finished building, including operation and maintenance (O&M) manuals, familiarisation training, tools, and spare parts; and

    (h) post-completion requirements such as maintenance services.

Instructions to tendering contractors and other information relating to the tender process (e.g. information required to be submitted with tender submission, site visits, confidentiality, etc) form part of the invitation documents, but will not form part of the contract documents. Therefore, these should be addressed in a separate document (e.g. Conditions of Tender).

The items to be considered when drafting the main contractor's preliminaries are included in Part A of Table 1 (Preliminaries (main contract)) at Part 3: Tabulated rules for measurement.

2.7.3.2 *Pricing schedule (quantification of main contractor's preliminaries):*

It is not possible for the quantity surveyor/cost manager to quantify the main contractor's preliminaries. This is because it is for the contractor to interpret the information provided as part of the tender invitation documentation. From the information provided, the main contractor will ascertain his method of working and the resources required to complete the building project, as well as identify any other cost items that are to be recovered.

The preliminaries bill is therefore to include a pricing schedule that simply lists the headings under which the main contractor is to price his or her preliminaries items. Templates for preliminaries pricing schedules (condensed and expanded versions) are included in Appendices B and C, respectively, of these rules.

The pricing schedule is a simple a list of cost centres incorporated in the bill of quantities in which the main contractor will insert his charges relating to preliminaries.

The pricing schedule for main contractors' preliminaries is divided into two main cost centres:

(1) employer's requirements; and

(2) main contractor's cost items.

The items that comprise these two cost centres are listed and defined in Part B of Table 1 (Preliminaries (main contract)) at Part 3: Tabulated rules for measurement of these rules.

Notwithstanding this, as part of the priced bill of quantities submitted by the main contractor, it is essential that the quantity surveyor/cost manager obtains a full and detailed breakdown that clearly identifies the items, shows how the price for each item has been calculated, and how the total price for preliminaries has been calculated.

Accordingly, as part of the conditions of tender, the quantity surveyor/cost manager should instruct the main contractor to return along with his or her tender a full and detailed breakdown that shows how the main contractor's total price for preliminaries has been calculated. It should be requested that the main contractor append this information to his priced bill of quantities. To ensure that the main contractor's detailed supporting calculations are presented in an easy-to-read and logical format, the main contractor should be instructed to ascertain the price for preliminaries in accordance with the rules of measurement for main contractor's preliminaries (refer to Part B (Pricing schedule) of Table 1 (Preliminaries (main contract)) at Part 3: Tabulated rules for measurement of these rules.

It is also essential that the quantity surveyor/cost manager makes it clear to the main contractor in the preliminaries bill and/or preliminaries pricing schedule that costs relating to items that are not specifically identified by the main contractor in his or her full and detailed breakdown will be deemed to have no cost implications or have been included elsewhere within his or her rates and prices.

2.7.4 Preliminaries (works package contract):

2.7.4.1 *Information:*

The descriptive part of the work package contract preliminaries is prepared in the same way as the main contract preliminaries. The items to be considered when drafting the work package contracts are included in Part A of Table 1 (Preliminaries (works package contract) at Part 3: Tabulated rules for measurement of these rules.

2.7.4.2 *Pricing schedule:*

For the same reasons as for main contractor's preliminaries, the preliminaries bill for a works package shall comprise a pricing schedule that lists the headings under which the work package contractor will insert his charges relating to preliminaries items. It is a simple schedule comprising a list of cost centres.

The pricing schedule for work package contract preliminaries is divided into two main cost centres:

(1)   employer's requirements; and

(2)   work package contractor's cost items.

The items that comprise these two cost centres are listed and defined in Part B of Table 1 (Preliminaries (works package contract)) at Part 3: Tabulated rules for measurement of these rules.

Again, as part of the priced bill of quantities submitted by the work package contractor, it is essential that the work package contractor is instructed to provide a full and detailed breakdown that clearly identifies the items, shows how the price for each item has been calculated, and how the total price for preliminaries has been calculated.

2.8   Measurement rules for building works

The rules for measuring and describing building items/components are set out in the tabulated rules of measurement rules for building works at Part 3 (Tabulated rules of measurement for building works) of these rules.

2.9   Non-measurable works

2.9.1   Provisional sums:

2.9.1.1   Where building components/items cannot be measured and described in accordance with the tabulated rules of measurement they shall be given as a 'provisional sum' and identified as either 'defined work' or 'undefined work' as appropriate.

2.9.1.2 A provisional sum for defined work is a sum provided for work that is not completely designed but for which the following information shall be provided:

(1)   the nature and construction of the work;

(2)   a statement of how and where the work is fixed to the building and what other work is to be fixed thereto;

(3)   a quantity or quantities which indicate the scope and extent of the work; and

(4)   any specific limitations identified.

2.9.1.3 Where provisional sums are given for defined work, the contractor will be deemed to have made due allowance in his or her programming, planning and pricing preliminaries.

2.9.1.4 Where any aspect of the information required by paragraph 2.9.1.2 above cannot be given, work shall be described as an 'undefined' provisional sum. Where provisional sums are given for undefined work, the contractor will be deemed not to have made any allowance in programming, planning and pricing preliminaries.

2.9.1.5 Any provisional sum given for defined work that does not comprise the information required under 2.9.1.2 above shall be construed as a provisional sum for undefined work; irrespective that it was given in the BQ as a provisional sum for defined work.

2.9.1.6 Provisional sums shall be exclusive of overheads and profit. Separate provision is to be made in the BQ for overheads and profit (refer to paragraph 2.11).

2.9.2   Contractor designed works:

2.9.2.1 Contractor designed works include any works that require the contractor to undertake its design, whether directly or via a subcontractor. The employer shall be deemed responsible for works not clearly identified as contractor designed works.

2.9.2.2 Contractor designed work is sometimes referred to as the 'Contractor Designed Portion (CDP)'.

2.9.2.3 Where the contractor is required to take responsibility for the design of discrete parts of the building, such as piled foundations, windows, pre-cast concrete components, roof trusses and/or mechanical and electrical engineering services, the work items shall be identified as 'contractor designed works'.

2.9.2.4 The method of quantifying contractor designed work is dependent on the nature of the work.

2.9.2.5 Where contractor designed works can be measured and described in accordance with the tabulated rules of measurement (e.g. windows and pre-cast concrete components), the performance objectives or criteria that the contractor will be required to meet are to be clearly defined by way of a preamble to the work items that comprise the contractor designed works. Detailed documents defining the performance objectives and/or criteria to be met are to be incorporated as an annex to the bill of quantities (BQ) and clearly cross-referenced in the preamble.

2.9.2.6 Where contractor designed works comprise a complete element or works package (e.g. the entire mechanical and electrical engineering services for the building), the works are to be measured and described as one or more item. The number of items is at the discretion of the quantity surveyor/cost manager but must be sufficient to provide an analysis of the price of the contractor designed works. In the case of elemental BQ, the basis of analysis will be the elements defined in *NRM 1: Order of cost estimating and cost planning for capital building works* (see Figure 2.2). Irrespective of the structure of the analysis, it is essential that the quantity surveyor/cost manager obtains a full and detailed breakdown that clearly shows how the contractor has calculated his price for each item in the analysis.

Figure 2.2: Price analysis for contractor designed works (based on group element 5 (Services) of the RICS new rules of measurement: Order of cost estimating and cost planning for capital building works (NRM 1))

5.1	Sanitary installations	£236,000.00
5.2	Services equipment	£199,500.00
5.3	Disposal installations	£99,000.00
5.4	Water installations	£142,000.00
5.5	Heat source	£88,000.00
5.6	Space heating and air conditioning	£396,000.00
5.7	Ventilation	£345,500.00
5.8	Electrical installations	£458,000.00
5.9	Fuel installations	£163,000.00
5.10	Lift and conveyor installations	£689,000.00
5.11	Fire and lightning protection	£222,300.00
5.12	Communication, security and control systems	£181,500.00
5.13	Specialist installations	£148,600.00
5.14	Builder's work in connection with services	£59,600.00
	Total (carried to main summary):	3,428,000.00

2.9.2.7 The quantity surveyor/cost manager is to obtain details of performance objectives and/or criteria from the relevant design consultant (refer to paragraph 2.14.3.2(2) below).

2.9.2.8 In addition to all costs referred to in paragraph 3.3.3.13 (in Part 3: Tabulated rules of measurement for building works) of these rules, contractor designed works shall be deemed to include all costs in connection with design, design management, and design and construction risks in connection with contractor designed works. Moreover, the contractor will be deemed to have made due allowance in his programming and planning for all design works in connection with contractor designed works.

2.9.3 Risks:

The method of dealing with the employer's residual risks when preparing bill of quantities (BQ) is addressed in paragraph 2.10.

2.9.4 Works to be carried out by statutory undertakers

2.9.4.1 Works that are required to be carried out by a statutory undertaker are to be given as a 'provisional sum', with the scope of works to be executed by the statutory undertaker described.

2.9.4.2 The contractor is to be deemed to have made due allowance in his programming, planning and pricing preliminaries for all general attendance on statutory undertakers.

2.9.4.3 Provisional sums for work to be carried out by statutory undertakers are to be exclusive of overheads and profit. Separate provision shall be made in the BQ for overheads and profit (refer to paragraph 2.11 for further information).

2.10 Risks

2.10.1 Risks generally:

2.10.1.1 Every building project involves risks and the proper management of risk saves time and money. Risks can occur at any point of a building project and it is essential that they are identified, assessed, monitored and controlled appropriately and effectively.

2.10.1.2 At the time of preparing a bill of quantities, a quantified schedule of works, or other quantity-based documents, whether for a complete building project or discrete works package, there will still be a number of risks remaining to be managed by the employer and his or her project team – this is called the employer's residual risk exposure (or residual risks). A risk response should only be decided after its possible causes and effects have been considered and fully understood. It will take the form of one or more of the following:

- risk transfer to the contractor;
- risk sharing by both employer and contractor; or
- risk retention by the employer.

2.10.1.3 Risks that can be designed out or avoided should have been addressed by this stage of the design development process. However, if time does not permit these risks to be designed out or properly dealt with, they will need to be dealt with using one of the above risk response strategies.

2.10.2 Risk transfer to the contractor:

2.10.2.1 The object of transferring risk is to pass the responsibility to another party able to control it better. If the risk materialises, the consequences are carried by the other party.

2.10.2.2 Whenever a risk is transferred there is usually a premium to be paid (effectively the contractor's valuation of the cost of the risk). Risk transfer will usually give the employer price/cost certainty for that aspect of the works. However, in return for price/cost certainty, the employer is required to pay the risk premium to the contractor, irrespective of whether the risk transferred does or does not materialise.

2.10.2.3 Risks that the contractor is required to manage, should they materialise, are to be fully described so that it is transparent what risk the contractor is required to manage, and what the extent of services and/or works the employer is paying for. Risks to be transferred to the contractor are to be listed in the BQ under the heading 'schedule of construction risks'. A template for a 'schedule of construction risks' is in Appendix F of these rules.

2.10.2.4 The contractor will be deemed to have made due allowance in his risk allowances for programming, planning and pricing preliminaries.

2.10.2.5 Risk allowances inserted by the contractor shall be exclusive of overheads and profit. Separate provision should be made in the BQ for overheads and profit (refer to paragraph 2.11 for further information).

2.10.3 Risk sharing by both employer and contractor:

2.10.3.1 Risk sharing occurs when a risk is not entirely transferred and some elements of it are retained by the employer. It is important that both the employer and the contractor know the value of the portion of the risk for which they are responsible. The objective should be to improve control and to reduce or limit the cost of the risk to the employer, should it materialise.

2.10.3.2 The method of dealing with risks that are to be shared by both employer and contractor will normally be dealt with using 'provisional quantities' with the pricing risk being taken by the contractor and the quantification risk being taken by the employer.

2.10.4   Risk retention by the employer:

2.10.4.1 Where risks are to be retained by the employer, the applicable risk allowances included in the cost plan will be retained and managed by the employer or, if empowered by the employer, the project team.

2.10.4.2 Before deciding to retain a particular risk, the employer might wish to find out what the premium will be if the contractor were to be paid for resolving the risk should it materialise. The employer can then decide whether or not to pay a premium for a defined scope of work. If the employer is content to pay a premium for transferring the risk, it is dealt with as a risk transfer in accordance with paragraph 2.10.2 above.

2.10.4.3 Risks retained by an employer are not necessarily controllable.

2.11 Overheads and profit

2.11.1 Provision shall be made in the bill of quantities (BQ) for the contractor to apply their required percentage addition for overheads and profit on the following:

(1)   preliminaries;

(2)   measured work, including contractor designed works;

(3)   risk allowances;

(4)   work resulting from the expenditure of provisional sums (i.e. defined provisional sums, undefined provisional sums and works to be undertaken by statutory undertakers).

2.11.2 When required, overheads and profit can be treated as two separate cost items; namely, 'overheads' and 'profit'.

2.12 Credits

2.12.1 Credits are normally only applicable where the building project comprises the refurbishment or rehabilitation of an existing building, or demolition works. Provision for such provides the employer an opportunity to seek credits for old building materials; components and items; mechanical and electrical plant; and fittings, furnishings and equipment which arise from the stripping out or demolition works and for which the employer is content to pass ownership to the contractor for reuse.

2.12.2 Credits can be based on a pre-prepared list of items, which is incorporated in the BQ and the contractor invited to insert the amount of credit he will give for each item. Alternatively, the contractor can be invited to list items for which he is willing to offer a credit and the amount of credit he will give for each item.

2.12.3 A template for 'credits' is provided in Appendix F of these rules.

2.13 Other considerations

2.13.1   Price fluctuations:

2.13.1.1 The cost to the contractor of labour and materials etc. used in the works will alter during the contract period (i.e. they will be subject to price fluctuations). It might fall but, more usually, it will rise. The risk of fluctuating prices can be dealt with as follows:

(1)   contractor to price the risk (a fixed or firm price contract); or

(2)   allow provision for contractor to recover full or limited fluctuations on certain prices (a fluctuating price contract).

2.13.1.2 Most standard forms of contract conditions allow for either alternative to be used by providing clauses which may be included or deleted.

(1)   *Fixed price contracts:*

These are contracts in which the price of labour, materials and plant is not subject to fluctuations. Fixed price contracts are sometimes referred to as 'fixed price lump sum contracts', 'firm price contracts', or 'firm price lump sum contracts'.

In the absence of any provision in the contract, or where the provision for recovering of price fluctuations has been deleted, the contractor will be required to take the risk (i.e. price the risk) of price fluctuations during the contract period. In order to cover himself, the contractor will make an estimate of the likely increase in costs and include this in his tender price.

Where there is no provision for recovering of price fluctuations, separate provision is to be incorporated in the bill of quantities for the contractor to tender his fixed price adjustment for pricing the risk. Such provision is to be referred to as either the 'main contractor's fixed price adjustment' or the 'work package contractor's fixed price adjustment', whichever is applicable.

When preparing bill of quantities, the quantity surveyor/cost manager shall ensure that no contract conditions relating to the recovery of price fluctuations exist.

(2)   *Fluctuating price contracts:*

These are contracts in which adjustment is allowed for fluctuations in the prices of labour and materials etc. Various degrees of fluctuations are allowed under the provisions of standard contract conditions. The extent to which fluctuations are allowed will have a significant effect upon the contractor's tender price.

Where fluctuations are to be allowed, no provision for 'main contractor's fixed price adjustment' or the 'work package contractor's fixed price adjustment' is required.

2.13.2   Director's adjustment:

2.13.2.1 Separate provision is to be incorporated in the bill of quantities for the contractor to insert a 'director's adjustment'.

2.13.2.2 It is the responsibility of the contractor's directors, or other senior managers, to secure work for the company. Therefore, before submitting a tender price, the contractor's directors will undertake a commercial review of the project and the estimated price. This review might result in the contractor's directors requiring adjustment to the estimated price, referred to as a 'director's adjustment'. The director's adjustment will include adjustments for commercial matters such as financing charges, cash flow, opportunities and competition. This is a sum added to or omitted from the estimated price to arrive at a tender price.

2.13.3   Dayworks (provisional):

2.13.3.1 Daywork is a method of valuing work on the basis of time spent by the contractor's employees, the materials used and the plant employed.

2.13.3.2 If required, a schedule of dayworks is to be incorporated in the bill of quantities (BQ). The schedule of dayworks is to comprise a list of the various classifications of labour, estimates of the number of hours against each classification of labour, and estimated lump sums for materials and plant, for which daywork rates and percentage additions for overheads and profit are to be inserted by the contractor. A statement of the conditions under which the contractor will be paid for work executed on a daywork basis is to be given in either the preliminaries bill or schedule of dayworks.

2.13.3.3 The method of calculating labour time charge rates for work carried out in normal working hours (i.e. productive time) and work carried out outside of normal working hours (i.e. non-productive time) shall be defined in the schedule of dayworks. The definition of normal working hours shall be given in either the preliminaries bill or schedule of dayworks.

2.13.3.4 The total amount included for daywork by the contractor shall be omitted from the contract sum. The rates and percentage additions included in the BQ shall be used to calculate the price of extra works instructed, which are authorised to be valued on a daywork basis.

Note: The total price derived from the schedule of dayworks can be included within or excluded from the contract sum. When included, it is to be treated as a provisional sum. When excluded, it shall be clearly stated that the rates, prices and percentage adjustments tendered are included in the contract.

2.13.4 Value Added Tax (VAT):

Value added tax (VAT) shall be excluded from bill of quantities (BQ). Notwithstanding this, if required by the employer, provision for the contractor to provide a VAT assessment as part of his tender return can be incorporated in the form of tender.

2.14 Information requirements for measurement

2.14.1 The accuracy of bill of quantities (BQ) is dependent on the quality of the information supplied to the quantity surveyor/cost manager by the employer, designers and other project team members; the more information provided, the more reliable the outcome will be. Where little or no information is provided, the quantity surveyor/cost manager will need to seek decisions from the employer as to how the uncertainty is to be managed and procured (refer to paragraph 2.11).

2.14.2 To enable the quantity surveyor/cost manager to prepare a BQ, the information in paragraphs 2.14.3 to 2.14.6 inclusive (below) will be required.

2.14.3 Specification:

2.14.3.1 A specification defines what the employer wishes to buy and, consequently, what the contractor is expected to supply.

2.14.3.2 The two main types of specification used are:
- Prescriptive specifications
- Performance specifications.

(1) *Prescriptive specifications:*

This form of specification is required to enable a firm bill of quantities to be prepared. The function of a prescriptive specification is to prescribe the materials and workmanship required for a building project in as much detail as possible. Specific products and materials will be named, and the assembly of the building will be described and supported with drawn information and schedules (see paragraphs 2.14.4 and 2.14.5 below). Where materials are not named, reference will be made to published standards governing their composition (e.g. British standards or other country specific standards).

Where a prescriptive specification is used, the contractor will not carry any design responsibility.

(2) *Performance specifications:*

A performance specification describes the requirements of a product (e.g. windows), equipment (e.g. chiller plant), system or installation (e.g. mechanical and electrical installations) in terms of the performance objectives or criteria.

The main difference between a performance specification and a prescriptive specification is related to design responsibility. With a performance specification, the contractor is responsible for design development of the specification to meet the performance requirements.

The benefit to the employer is that design will not need to be advanced for performance-specified work before inviting tenders from contractors.

It should be noted that some standard forms of contract conditions do not include provision for contractor designed work. Therefore, care should be taken to ensure that the contract conditions used for the building project clearly transfer design responsibility for contractor designed work to the contractor (e.g. by incorporating supplementary contract conditions or other amendment to the standard forms of contract conditions used). Failure to do this will result in design responsibility remaining with the employer, even though the employer did not undertake the design.

Usually, the quantity surveyor/cost manager will be faced with a combination of both types of specification, which needs to be organised in the BQ (i.e. separated into measured works and non-measured works).

Insufficient or poorly described information can mislead contractors, resulting in contract variations and potential time-related and/or cost-related claims.

2.14.4 Drawn information:

2.14.4.1 Drawn information is required to describe the assembly of the building, as well as any temporary works. Drawings shall be to a suitable scale.

2.14.4.2 Required drawn information shall include:

- *General arrangement (GA) drawings*, comprising:

    - a block plan: this shall identify the site and locates the outlines of the building works in relation to the town plan or other wider context;

    - a site plan: this shall locate the position of the building works in relation to the setting out points, the means of access and the general layout of the site; and

    - plans, sections and elevations: these shall show the position occupied by the various spaces in a building and the general construction and location of the principal elements and components. The extent of elevations and sections shall be as appropriate to cover all major building zones.

    General arrangement drawings are sometimes called 'location' drawings.

- *Component drawings*: these shall show the information necessary for manufacture and assembly of a component; including key details/interfaces (e.g. interface between curtain walling system and structure, balconies and the like).

    Component drawings are sometimes called 'detail' or 'assembly' drawings.

- *Schematic drawings*: these show how something works and the relation between the parts (e.g. the wiring of an electrical system).

- *Record drawings*: these are a set of drawings that depict the actual as-built conditions of an existing building or structure, including mechanical and electrical engineering services installed. These are required for building projects involving the refurbishment or the demolition (partial or complete) of an existing building or structure.

    Record drawings are sometimes called 'existing', 'as-built' or 'as-installed' drawings.

2.14.4.3 Specific requirements for drawn information are further defined in the tabulated rules of measurement in Part 3 of these rules.

2.14.5 Schedules:

2.14.5.1 Schedules which provide the information required by the tabulated rules shall be deemed to be drawings. Schedules include:

- Room data sheets (including co-ordinated mechanical and electrical services engineering data sheets);

- Door schedules, including ironmongery;

- Window schedules, including ironmongery;

- Reinforcement (bar bending) schedules

- Landscaping and planting schedule (for internal and external works);

- Drainage schedules;

- Fittings, furnishings and equipment schedules;

- Luminaires schedules;

- Control schedules for mechanical and electrical engineering services;

- Primary mechanical and electrical plant and equipment schedules;

- Duties, outputs and sizes of primary mechanical and electrical plant and equipment;

- Builder's work in connection with mechanical and electrical engineering services;
- Other scheduled information necessary to specify the works.

The above list is not meant to be definitive or exhaustive, but simply a guide.

2.14.6 Reports and other information:

Reports and/or other information required for the preparation of bill of quantities (BQ) will be dependent on the nature of the building project. However, such documents normally include some or all of the following:

- drawings showing the site boundary and all known site constraints and restrictions, including the means of access, restrictive covenants, party walls, rights of light issues, and rights of access;
- a statement of, and drawing(s) detailing, phasing requirements;
- a statement of, and drawing(s) detailing, construction sequencing requirements;
- details of residual design development and construction risks (i.e. risk register or risk log);
- a schedule of gross external areas (GEA), gross internal floor areas (GIFA), net internal areas (NIA – i.e. usable area for shops, supermarkets and offices) and site area (SA).
- site survey reports, including archaeological survey, ecological survey, invasive plant growth survey, etc.;
- details of wildlife, including protected species, protection measures;
- geotechnical and report(s) describing the intrusive ground and groundwater investigations completed, together with the results (e.g. the results of trial pits, auger holes, window samplers, boreholes, cone penetration tests (CPTs), and standard penetration tests (SPTs));
- environmental report(s) describing the sampling and analysis of soils, together with the results, giving information about the soil, ground water, and gases;
- details of any other facilitating works (e.g. soil stabilisation measures);
- remediation plan describing the method of dealing with contaminated materials and/or invasive plant growth (e.g. Japanese knotweed and giant hogweed), including requirements for post-remediation validation sampling;
- refurbishment and demolition survey reports, providing details of any asbestos containing materials (ACMs) and/or other hazardous materials, together with the scope of removal or encapsulation works required to be undertaken as part of the building works;
- details of any other facilitating works (e.g. soil stabilisation measures);
- details of any party wall awards or other agreements with adjoining owners and statutory undertakers, specifically detailing any requirements of the award to which the contractor should comply to ensure that the employer does not breach any agreement;
- temporary works methodology, drawings and sketches;
- details of which condition of contract is to be used for the building project;
- details of any planning conditions or informatives that the contractor is required to comply with;
- the employer's requirements in respect of insurances;
- the employer's requirements for the contractor to collect and report cost data to support claims for capital allowances, grants, value added tax (VAT) recovery, and other tax incentives;
- the employer's policy documents, which the contractor will be required to comply with (e.g. site rules and regulations, environmental, corporate social responsibilities, and health and safety policies);
- hoarding requirements, including design where this is an employer's requirement;
- details of the employer's post completion requirements (e.g. operation and maintenance of completed building (i.e. works and services)); and
- all other information necessary to construct the preliminaries bill for the building project or work package (refer to Work Section I (Preliminaries) at Part 3 (Tabulated rules of measurement for building works) of these rules).

The above list is not meant to be definitive or exhaustive, but merely a guide. It is intended to be used by the quantity surveyor/cost manager to assist in identifying the types of reports and other information required to prepare a robust BQ.

2.15 Codification of bill of quantities

2.15.1 Planning the bill of quantities breakdown structures:

2.15.1.1 Before attempting to prepare a bill of quantities (BQ) for a building project, the composition of the building project needs to be determined and the structure of the BQ planned. The composition of a BQ can be viewed as a work breakdown structure (WBS). This is a tree structure which can be used to define and divide a building project into key facets. It is developed by starting with the end objective (i.e. WBS Level 0 – the entire building project) and successively subdividing it into the main components and sub-components that make up the entire building project – providing a hierarchical breakdown. What is more, a WBS initiates the development of the cost breakdown structure (CBS), which can be used to allocate costs to every facet of the building project at each level of the WBS. Together, the WBS and CBS provide a frame of reference for the cost management of a building project during the construction phase (i.e. post contract). In the context of BQ, the WBS is referred to as BQ breakdown structure (BQBS).

2.15.1.2 There are three principal breakdown structures for BQ. They are:

(1) *Elemental*:

Measurement and description is done by group elements; following the logic arrangement for elemental cost planning defined in *NRM 1: Order of cost estimating and cost planning for capital building works*. Each group element forms a separate section of the BQ, irrespective of the order of work sections in *NRM 2: Detailed measurement for building works*. Group elements are sub-divided through the use of elements, which are further sub-divided by sub-elements.

(2) *Work section*:

Measurement and description is divided into the work sections defined in *NRM 2: Detailed measurement for building works* (refer to the tabulated rules in Part 3 of these rules).

(3) *Work package*:

Measurement and description is divided into employer or contractor defined work packages. Works packages can be based on either a specific-trade package or a single package comprising a number of different trades.

2.15.1.3 Amplification of these three principal BQ breakdown structures is given in paragraph A.1 (Bill of quantities breakdown structures) at Appendix A (Guidance on the preparation of bill of quantities) of these rules.

2.15.2 Codification of bill of quantities:

2.15.2.1 The work breakdown structure (WBS) for the building project will have been initiated by the quantity surveyor/cost manager when preparing the initial order of cost estimates and developed during the formal cost planning stages (i.e. the cost plan breakdown structure). As part of this process, the cost breakdown structure (CBS) will have evolved. The key benefit of the cost plan breakdown structure is the ability to uniquely identify by a code all group elements, elements, sub-elements, and components (i.e. building components/items) in a numerical and logical manner – providing a codification framework for the cost management, control and reporting of costs. With a unique code, all building components/items can be linked to components, components to sub-elements, sub-elements to elements, elements to group elements, and group elements to the cost limit (i.e. the total estimated cost of the building project). This makes it easier to retrieve, manage and restructure information (i.e. costs and building components/items). Details of the codification framework, including the numbering logic, advocated for cost planning can be found in *NRM 1: Order of cost estimating and cost planning for capital building works*.

2.15.2.2 For the quantity surveyor/cost manager to manage the cost plan during the procurement and construction phases of the building project (i.e. by reconciling tender prices and project outturn costs against the cost plan), the codification framework used for cost planning must be used as the basis for the codification of building components/items and components in the bill of quantities (BQ).

2.15.3   The coding system:

2.15.3.1 The coding system used for a bill of quantities (BQ) will be dependant on the BQ breakdown structure used. That is:

- Elemental breakdown structure;
- Work sectional breakdown structure; or
- Work package breakdown structure.

(1)   *Elemental breakdown structure*

When preparing elemental BQ, the BQ breakdown structure is based on the group elements defined by *NRM 1: Order of cost estimating and cost planning for capital building works* (i.e. facilitating works; substructure; superstructure; internal finishes; fittings, furnishings and equipment; services; prefabricated buildings and building units, work to existing buildings, and external works).

The identification numbers used to formulate codes for cost planning are described in *NRM 1: Order of cost estimating and cost planning for capital building works*. The coding system advocated by NRM 1 is numeric. However, both alpha and numeric codes can be used (letters and numbers). It is recommended that the same approach be used in codifying building components/items (i.e. components and sub-components) in bill of quantities (BQ).

For practical purposes five to six levels of code are considered sufficient in cost planning to achieve the desired level of pre-contract cost control of a building project. The main identification numbers levels are as follows:

- *Level 0: Project number – most building projects will be given a project number, together with a project title or name, to distinguish them from all other projects the company might be working on.*

- *Level 1: Cost plan number – where a building project comprises more than one building or facet, a discrete cost plan will most likely be prepared for each building and key facet; culminating in a 'summary cost plan'. Therefore, an identification number will be required to distinguish cost plans. This code will not be required for a single cost plan.*

The identification numbers for Level 2, Level 3 and Level 4 are pre-defined by *NRM 1: Cost estimating and cost planning for building works*.

- *Level 2: Group element – identification number pre-defined.*

- *Level 3: Element – identification number pre-defined.*

- *Level 4: Sub-element – identification number pre-defined.*

- *Level 5: Component – user defined (building components/items).*

Because building components/items are described and quantified in greater detail in BQ than those for cost planning, a user defined level 6 identification numbers will need to be introduced for each sub-component of a component that is to be measured in accordance with *NRM 2: Detailed measurement for building works*. The way in which a code can be expanded to include a level 6 identification number for sub-components is shown in the following example.

Figure 2.3: Example of resultant codes used for codifying components and sub-components in elemental bill of quantities

Level	Description	Item	Identification numbers	Resultant codes
0	Project number		DPB27	
1	Bill number	Bill no. 3	3	
2	Group element/BQ number	Substructure	1	
3	Element	Foundations	1	
4	Sub-element	Piled foundations	2	
5	Component	Pile cap	1	
6	Sub-component	Excavation	1	DPB27-3.1.1.2.1.1
6	Sub-component	Disposal	2	DPB27-3.1.1.2.1.2
6	Sub-component	Concrete	3	DPB27-3.1.1.2.1.3
6	Sub-component	Reinforcement	4	DPB27-3.1.1.2.1.4
6	Sub-component	Formwork	5	DPB27-3.1.1.2.1.5

Note: It might not be necessary to prefix the code for components and sub-components with the project number throughout the BQ.

The resultant codes can be inserted in the right-hand column of the bill paper or in brackets after the bill description.

(2) *Work sectional breakdown structure*

Where a work sectional breakdown structure is used to construct the bill of quantities (BQ), the work sections will be those defined in the work sections in *NRM 2: Detailed measurement for building works* (refer to the tabulated rules in Part 3 of this document). However, for the purposes of cost management and cost control, it is essential that the work sectional breakdown structure can be easily reconciled with the original cost plan breakdown structure.

The method recommended by the rules involves the provision of a secondary code which acts as a suffix to the primary code used for BQ based on an elemental breakdown structure described in sub-paragraph a) above. Examples of suffix codes are illustrated in Figure 2.4 below.

Figure 2.4: Example of suffix codes used for codifying work sections in work sectional BQ breakdown structure

Serial no.	Work Section	Suffix
1	Preliminaries (main contract)	/01
1	Preliminaries (works package contract)	/01
2	Off-site manufactured materials, components and buildings	/02
3	Demolitions	/03
4	Alterations, repairs and conservation	/04
5	Excavating and filling	/05
6	Ground remediation and soil stabilisation	/06
7	Piling	/07
8	Underpinning	/08
9	Diaphragm walls and embedded retaining walls	/09
10	Crib walls, gabions and reinforced earth	/10
11	'In-situ' concrete works	/11

Serial no.	Work Section	Suffix
12	Precast/composite concrete	/12
13	Precast concrete	/13
14	Masonry	/14
15	Structural metalwork	/15
16	Carpentry	/16
17	Sheet roof coverings	/17
18	Tile and slate roof and wall coverings	/18
19	Waterproofing	/19
20	Proprietary linings and partitions	/20
21	Cladding and covering	/21
22	General joinery	/20
23	Windows, screens and lights	/21
24	Doors, shutters and hatches	/20
25	Stairs, walkways and balustrades	/21
26	Metalwork	/20
27	Glazing	/21
28	Floor, wall, ceiling and roof finishings	/20
29	Decoration	/21
30	Suspended ceilings	/30
31	Insulation, fire stopping and fire protection	/31
32	Furniture, fittings and equipment	/32
33	Drainage above ground	/33
34	Drainage below ground	/34
35	Site works	/35
36	Fencing	/36
37	Soft landscaping	/37
38	Mechanical services	/38
39	Electrical services	/39
40	Transportation	/40
41	Builder's work in connection with mechanical, electrical and transportation installations	/41

Notes:

1 Both alpha and numeric codes can be used (letters and numbers).

2 As with elemental BQ, it is not be necessary to prefix the code for components and sub-components with the project number throughout the BQ.

Using the examples of sub-components given in Figure 2.3 above, the resulting codes for a work sectional breakdown structure will be:

- Excavation   DPB27-3.1.1.2.1.1/04
- Disposal   DPB27-3.1.1.2.1.2/04
- Concrete   DPB27-3.1.1.2.1.3/04
- Reinforcement   DPB27-3.1.1.2.1.4/04
- Formwork   DPB27-3.1.1.2.1.5/04

(3)   *Work package breakdown structure*

Both *NRM 1: Order of cost estimating and cost planning for capital building works* and *NRM 2: Detailed measurement for building works* recognise that cost plans will need to be restructured from elements to works packages for the purposes procurement. However, they make no attempt at standardising works packages. This is because the content of work packages is likely to be different from one building project to another – with the content of work packages often based on the perception of risk to those ultimately liable for the construction works. For example, for one building contract it might be deemed appropriate to have all concrete work carried out under by a single subcontractor. Whereas for another building contract, because of the perceived risks associated with the drainage passing through the ground floor construction, it is considered more appropriate to include the construction of pile caps, ground beams, base slab and below ground drainage in the works package for groundworks.

For that reason, the number and content of work packages need to be carefully planned by the cost manager/quantity surveyor before commencing the preparation of the bill of quantities. Once the work package breakdown structure has been established, the provision of a secondary code which acts as a suffix to the primary code (i.e. that used for BQ based on an elemental breakdown structure described in sub-paragraph (a) above) can be applied.

Figure 2.5 provides typical example of part of a work package breakdown structure, to which suffix codes have been applied. An example of a full work package breakdown structure is given at Appendix G of these rules. The example is not meant to be definitive or exhaustive, but simply for guidance.

Figure 2.5: Typical example of suffix codes used for codifying work packages when a work package BQ breakdown structure is used

Serial no.	Work package title/content	Suffix
1	Preliminaries	/01
2	Intrusive investigations: Asbestos and other hazardous materials Geotechnical and environmental investigations Attendance on archaeological investigations	/02
	Preliminaries (main contract)	/01.2
3	Demolition works: Asbestos and other hazardous materials removal/treatment works Soft strip of building components and sub-components Soft strip of mechanical and electrical engineering services Demolition	/03
	Preliminaries (works package contract)	/01.2
4	Groundworks: Contaminated ground material removal Preparatory earthworks Excavation and earthworks, including basement excavation, earthwork support and disposal Temporary works – propping of existing basement retaining walls Below ground drainage Ground beams Pile caps Temporary works – piling mats/platforms Ground bearing base slab construction, including waterproofing Basement retaining wall structures, including waterproofing	/04
	Preliminaries (works package contract)	/01.2
5	Piling: Piling works	/05
	Preliminaries (works package contract)	/01.2

Serial no.	Work package title/content	Suffix
6	Concrete works: Frame Upper floors, including roof structure Core and shear walls Staircases Preliminaries (works package contract)	/06 /01.2
7	Roof coverings and roof drainage	/07
8	External and internal structural walls	/08
9	Cladding	/09
10	Windows and external doors	/10
11	Mastic	/11
12	Non-structural walls and partitions	/12
13	Joinery	/13
14	Suspended ceilings	/14
15	Architectural metalwork	/15
16	Tiling	/16
17	Painting and decorating	/17
18	Floor coverings	/18
19	Fittings, furnishings and equipment	/19
20	Combined Mechanical and Electrical Engineering Services	/20
21	External works	/21

Notes:

1. Only work packages 1 to 6 inclusive have been expanded to illustrate the typical content of the work package.
2. As with elemental BQ, it is not be necessary to prefix the code for components and sub-components with the project number throughout the BQ.

Again, using the examples of sub-components given in Figure 2.3 above, the resulting codes for a work package breakdown structure will be:

- Excavation   DPB27-3.1.1.2.1.1/04
- Disposal   DPB27-3.1.1.2.1.2/04
- Concrete   DPB27-3.1.1.2.1.3/04
- Reinforcement   DPB27-3.1.1.2.1.4/04
- Formwork   DPB27-3.1.1.2.1.5/04

As all sub-components relate to the construction of pile caps, they will all be incorporated in the work package for groundworks and, as a consequence, all be given the same suffix.

2.16 Cost management/control

2.16.1 The main purpose of a bill of quantities (BQ) is to present a co-ordinated list of components/items, together with their identifying descriptions and quantities that encompass the building works so that the tendering contractors are able to prepare tenders efficiently and accurately. As well as assisting in ensuring parity of tendering. In addition, BQ provide a vital tool, which can be used by the quantity surveyor/cost manager to manage and control the costs of the building project. Cost management and control uses include:

- Pre-tender estimates;
- Post tender estimates;
- Cost planning;
- Pricing variations; and
- Interim valuations and payment.

2.16.2   Pre-tender estimate:

Pre-tender estimates are prepared immediately before calling the first tenders for construction. This is the final cost-check undertaken by the cost manager before tender bids for the building project, or any part of the building project, are obtained. When a bill of quantities (BQ) is the basis of obtaining a tender price, the pre-tender estimates will be based on the BQ.

2.16.3   Post-tender estimate:

A post-tender estimate is prepared at RIBA Work Stage H (tender action), or OGC Gateway 3C (investment decision), after all the construction tenders have been received and evaluated. It is based on the outcome of any post-tender negotiations, including the resolution of any tender qualifications and tender price adjustments. The post-tender estimate will include the actual known construction costs and any residual risks. The aim of this estimate is to corroborate the funding level required by the employer to complete the building project, including cost updates of project and design team fees, as well as other development and project costs, where they form part of the costs being managed by the cost manager. When reporting the outcome of the tendering process to the employer, the quantity surveyor/cost manager should include a summary of the post-tender estimate(s). The post-tender estimate should be fairly accurate because the uncertainties of market conditions have been removed. Post-tender estimates are used as the control estimate during construction.

2.16.4   Cost planning:

2.16.4.1 Cost planning is an iterative process, which is performed in steps of increasing detail as more design information becomes available. A cost plan provides both a work breakdown structure and a cost breakdown structure which, by codifying, can be used to redistribute works in elements to works packages for the purpose of procurement and cost control during the construction phase of the building project.

2.16.4.2 The third formal cost plan stage (completed at RIBA Work Stages E/F (Technical Design/Production Information) is based on technical designs, specifications and detailed information for construction. Formal cost plan 3 will provide the frame of reference for appraising tenders. It also provides the frame of reference for reconciling 'actual costs' against 'cost targets'. This is particularly beneficial where building works are being procured piecemeal (e.g. procuring discrete work packages as their design is completed). The proper use of the cost plan will allow the re-profiling of cost targets as necessary to ensure that the overall cost limit (i.e. the employer's authorised budget) is maintained by the project team.

2.16.5   Pricing variations:

The rates in a priced bill of quantities (BQ) provide a basis for the valuation of varied work. 'Pro-rata' and 'analogous' rates can also be ascertained from the base rates tendered to calculate the prices of other components not specifically described in the BQ.

2.16.6   Interim valuations and payment:

2.16.6.1 Many building projects require interim payments to be paid to the contractor. This is in order to relieve the contractor of the burden of financing the whole of the building works until completion; works which may take many months or years to complete. Within each contract there will be clauses which set out the administrative rules under which the quantity surveyor/cost manager, architect (or contract administrator or project manager, quantity surveyor/cost manager) employer and contractor must operate. In many contracts, while the completion and calculation of the value is important, the method and procedure of the interim valuations and payment which the contractor receives is equally important.

2.16.6.2 A priced bill of quantities (BQ) provides a comprehensive list of building components/items. Consequently, when a contract has been entered into, by assessing the building components/items in the BQ, the priced BQ can be used to ascertain periodic valuation of works properly executed in accordance with the provisions of the contract for the purpose of interim valuations and payment.

2.17 Analysis, collection and storage of cost data

2.17.1 Priced bill of quantities (BQ) make available one of the best sources of real-time cost data, which can be used by quantity surveyors/cost managers to provide expert cost advice on the likely cost of future building projects. Moreover, they afford a complete cost model in a single document.

2.17.2 The cost data provided in a BQ can be retrieved, analysed, stored and reprocessed in various ways (e.g. as distinct rates, detailed elemental cost analyses, element unit rates (EUR), cost/m^2 of gross internal floor area, and/or functional unit rates) for use in order of cost estimates and cost plans. It can also be used for benchmarking purposes.

2.17.3 *NRM 1: Order of cost estimating and cost planning for capital building works* can be used as a basis for measuring element unit quantities (EUQ) for the purpose of preparing detailed cost analyses of building projects.

Part 3: Tabulated rules of measurement for building works

Part 3: Tabulated rules of measurement for building works

3.1 Introduction

3.1.1 Part 3 of the rules comprise:

(1) the information and requirements for main contractor's and work package contractors' preliminaries, together with the rules for preparing the preliminaries pricing schedule; and

(2) the rules of measurement for building components/items.

3.1.2 The use of the tabulated rules is also explained.

3.1.3 Bill of quantities (BQ) are to fully describe and accurately represent the quantity and quality of the works to be carried out. More detail than is required by these rules should be given where necessary to define the precise nature and extent of the required work.

3.2 Use of tabulated rules of measurement for building works

3.2.1 General

3.2.1.1 The rules of measurement for building works are set out in tables. The tables are divided into two categories, namely those dealing with:

- preliminaries; and
- measurement of building components/items.

3.2.1.2 The rules are written in the present tense.

3.2.1.3 The symbol '/' used between two or more units of measurement or within text, means 'or'.

3.2.1.4 Horizontal lines divide the tables and rules into zones to which different rules apply.

3.2.1.5 Where units of measurement or rules are separated by a broken line (- - - - -) this denotes a choice of units or choice of ways of measuring the work. The method chosen shall be the best to suit the particular situation.

3.2.1.6 The use of a hyphen (-) or the phrase 'to' between two dimensions in these tables or in a bill of quantities means a range of dimensions exceeding the first dimension stated but not exceeding the second.

3.2.2 Tables: preliminaries

3.2.2.1 Work Section 1 comprises the rules for describing and quantifying preliminaries. It is divided into two sub-sections as follows:

- preliminaries (main contract); and
- preliminaries (work package contract).

3.2.2.2 Both sub-sections are sub-divided into parts as follows:

- Part A: Information and requirements (i.e. dealing with the descriptive part of the preliminaries); and
- Part B: Pricing schedule (i.e. provides the basis of a pricing document for preliminaries).

3.2.2.3 The tables for information and requirements for both preliminaries (main contract) and preliminaries (work package contract) are structured as follows:

    (1) Information and requirements are specified under a number of headings, which are given above each table (i.e. project particulars; drawings and other documents; and the site and existing buildings);

    (2) The left hand column (sub-heading 1) lists the preliminaries items to be considered under each main heading;

    (3) The second column (sub-heading 2) lists the sub-items to be considered under each sub-heading;

    (4) The third column (information requirements) lists the information which shall be included within the preliminaries descriptions; and

    (5) The fourth column (supplementary information/notes) lists supplementary information that might need to be included within the preliminaries descriptions, as well as providing additional guidance on the drafting of preliminaries statements.

3.2.2.4 The pricing schedule tables for both preliminaries (main contract) and preliminaries (work package contract) are structured as follows:

    (1) The contractor's pricing of preliminaries are captured under a number of headings, which are given above each table (i.e. project particulars; drawings and other documents; and the site and existing buildings);

    (2) The left hand column (component) lists the preliminaries items to be considered under each main heading;

    (3) The second column (included/notes on pricing) lists the sub-items which form part of each item;

    (4) The third column (unit) lists the unit of measurement for preliminaries items;

    (5) The fourth column (pricing method) stipulates if the component is a 'fixed charge', a 'time-related charge', or a combination of both; and

    (6) The fifth column (excluded) describes the items excluded from a component. Where exclusions are stated, cross references to the appropriate component is given.

3.2.3   Tables: building components/items

3.2.3.1   Work sections 2 to 41 comprise the rules of measurement for building components/items. They are as follows:

No.	Work Section:
2	Off-site manufactured materials, components and buildings;
3	Demolitions;
4	Alterations, repairs and conservation;
5	Excavating and filling;
6	Ground remediation and soil stabilisation;
7	Piling;
8	Underpinning;
9	Diaphragm walls and embedded retaining walls;
10	Crib walls, gabions and reinforced earth;
11	In-situ concrete works;
12	Precast/composite concrete;
13	Precast concrete;
14	Masonry;
15	Structural metalwork;
16	Carpentry;
17	Sheet roof coverings;
18	Tile and slate roof and wall coverings;
19	Waterproofing;
20	Proprietary linings and partitions;
21	Cladding and covering;
22	General joinery;
23	Windows, screens and lights;
24	Doors, shutters and hatches;
25	Stairs, walkways and balustrades;
26	Metalwork;
27	Glazing;
28	Floor, wall, ceiling and roof finishings;
29	Decoration;
30	Suspended ceilings;
31	Insulation, fire stopping and fire protection;
32	Furniture, fittings and equipment;
33	Drainage above ground;
34	Drainage below ground;
35	Site works;
36	Fencing;
37	Soft landscaping;
38	Mechanical services;
39	Electrical services;
40	Transportation; and
41	Builder's work in connection with mechanical, electrical and transportation installations.

3.2.3.2   Each table is structured as follows:

(1)   The title of the work section is given in the heading;

(2)   The first two rows set out the:

(a)   drawn information required in respect of each work section to enable measurement and shall accompany the bill of quantities when issued,

(b)   mandatory information that is to be provided in each work section,

(c)   minimum information that shall be shown on the drawings or any other document that accompany each work section, and

    (d)   works and materials that not measured, but are deemed to be included in the building components/items measured in each work section;

  (3)   The left hand column (item) lists the building components/items commonly encountered in building works;

  (4)   The second column (unit) lists the unit of measurement for building components/items;

  (5)   The third column (level 1) lists the information, including any dimension requirements, that shall be included in the description of the building components/items;

  (6)   The fourth column (level 2) lists the supporting information, including any additional dimension requirements, which shall be included in the description of the building components/items;

  (7)   The fifth column (level 3) lists the further supporting information, including any additional dimension requirements, which shall be included in the description of the building components/items; and

  (8)   The sixth column (notes, comments and glossary) explains what work is deemed to be included in specific building components/items, clarifies the approach to quantification and description of building components/items, and contains definitions of specific terms and phrases used in connection with the building components/items.

The building components/items listed in the tables comprise those commonly encountered in building works; the lists are not intended to be exhaustive.

3.3   Measurement rules for building works

3.3.1   BQ shall fully describe and accurately represent the quantity and quality of the works to be carried out. Where necessary, more detail than is required by these rules shall be given in order to define the precise nature and extent of the required work.

3.3.2   **Quantities:**

The rules for quantifying building components/items are as follows:

  (1)   *Measurement and billing:*

    (a)   Measure work net as fixed in position unless otherwise stated.

    (b)   Net quantity measured shall be deemed to include all additional material required for laps, joints, seams and the like, as well as any waste material.

    (c)   Curved work shall be measured on the centre line of the material unless otherwise stated.

    (d)   Dimensions shall be measured to the nearest 10mm. 5mm and over shall be regarded as 10mm and less than 5mm shall be disregarded.

    (e)   Except for quantities measured in tonnes (t), quantities shall be given to the nearest whole number. Quantities less than one unit shall be given as one unit. Quantities measured in tonnes (t) shall be given to two decimal places.

  (2)   *Voids:*

    (a)   Unless otherwise stated, minimum deductions for voids refer only to openings or wants within the boundaries of the measured work.

    (b)   Always deduct openings or wants at the boundaries of measured areas, irrespective of size.

    (c)   Do not measure separate items for widths not exceeding a stated limit where these widths are caused by voids.

3.3.3   **Descriptions:**

3.3.3.1   Each work section of a bill of quantities shall begin with a heading and a description stating the nature and location of the work.

3.3.3.2 Headings for groups of building components/items (i.e. components and sub-components) in a bill of quantities shall be read as part of the descriptions of the items to which the headings apply.

3.3.3.3 Descriptions shall state the building components/items being measured (taken from the first column of the tabulated rules) and include all Level 1, 2 and 3 information (taken from the third, fourth and fifth columns respectively) applicable to that item. Where applicable, the relevant information from column five shall be included in the description.

3.3.3.4 Unless specifically stated otherwise in the bill of quantities or in these rules, descriptions for building components/items shall include the:

(1)   type and quality of the material;

(2)   critical dimension(s) of the material(s);

(3)   method of fixing, installing or incorporating the goods or materials into the work where not at the discretion of the contractor; and

(4)   nature or type of background.

3.3.3.5 Where the nature or type of background is required to be identified, the building components/items description shall state one of the following:

(1)   timber (the term includes all types of hard and soft building boards);

(2)   plastics;

(3)   masonry (the term includes brick, concrete, block, natural and reconstituted stone);

(4)   metal, of any type;

(5)   metal-faced timber or plastics; and

(6)   vulnerable materials (the term includes glass, marble, mosaic, ceramics, tiled finishes, material finishes and the like).

3.3.3.6 Dimensions given as part of the description shall be:

(1)   stated in the sequence: length, width and height. Where ambiguity could arise, the dimensions shall be identified in the description;

(2)   the finished lengths, widths and heights specified or shown on the drawings with no allowance made for overlaps, scarcements and the like.

3.3.3.7 Thicknesses given as part of the description shall be the finished thickness of the material after compaction and shall exclude the thickness of adhesives and or bedding materials unless otherwise stated.

3.3.3.8 The use of a hyphen (-) or the phrase 'to' between two dimensions in a description shall mean a range of dimensions exceeding the first dimension stated but not exceeding the second.

3.3.3.9 Where the rules require work to be described as 'curved' with the radius stated, details shall be given of the curved work, including if concave or convex, if conical or spherical, if to more than one radius, and shall state the radius or radii. The radius shall be the mean radius measured to the centre line of the material unless otherwise stated.

3.3.3.10 The information required by these rules may be given by a precise and unique cross-reference to another document (e.g. to a specification or to a catalogue).

3.3.3.11 Where other components and sub-components are referred to in other documents (e.g. a specification states that vinyl sheet flooring is to be laid on a plywood lining), each component and or sub-component shall be measured and described separately (i.e. both the vinyl sheet flooring and the plywood lining are to be measured as separate items).

3.3.3.12 Notwithstanding the requirements of paragraph 3.3.3.11 above, separate components or sub-components may be combined to form single composite building components/items. In such cases, the description of the composite building components/items shall clearly state what is included and how each component and or sub-component is to be incorporated. Any component, sub-component or other element of the work not clearly included in the description shall be deemed not to be included as part of the composite building components/items.

3.3.3.13 Unless specifically stated otherwise in the BQ or in these rules, each building component/item shall be deemed to include the following:

 (1) labour and all costs in connection therewith;

 (2) materials and goods together with all costs in connection therewith;

 (3) assembling, installing, erecting, fixing or fitting materials or goods in position;

 (4) plant and all costs in connection therewith;

 (5) waste of goods or materials;

 (6) all rough and fair cutting unless specially stated otherwise;

 (7) establishment charges; and

 (8) cost of compliance with all legislation in connection with the work measured including health and safety, disposal of waste and the like.

3.3.4 Work of special types:

3.3.4.1 Work of each of the following special types shall be separately identified. Work of special types includes:

 (1) Work to existing buildings: such work is defined as work on, in or immediately under work existing before the current building project. Specific details pertaining to work carried out to an existing building shall be given at the start of each applicable work section.

 (2) Work carried out and subsequently removed: specific details pertaining to work that is to be carried out and subsequently removed shall be given at the start of each applicable work section.

 (3) Work outside the curtilage of the site: specific details pertaining to work to be executed outside the curtilage of the site shall be given at the start of each applicable work section.

 (4) Work carried out in extraordinary conditions, including:

  (a) in, on or under water, stating whether river, canal, lake or sea and, where applicable, stating the mean spring levels of high and low water;

  (b) in tidal conditions;

  (c) underground, stating mean depth;

  (d) in compressed air, stating the pressure and means of entry and exit; and

  (e) in other types of extraordinary conditions.

3.3.4.2 Specific details pertaining to work carried out in each condition shall be given at the start of each applicable work section.

3.3.4.3 The additional rules for special types of work shall be read in conjunction with the rules in the appropriate work sections.

3.3.4.4 Details of the additional preliminaries that are pertinent to the special types of work shall be given in the description, drawing attention to any specific requirements due to the nature of the work.

3.3.5 Measurable work not covered by the tabulated rules:

3.3.5.1 Building components/items not covered by the tabulated rules shall, if possible, be measured by rules for similar types of work. Rules of measurement adopted for such building components/items shall be clearly stated and fully defined in either the preliminaries or in the bill of quantities (against the building components/items or items to which the rule relates). Such rules shall, as far as possible, conform to those given in the tabulated rules of measurement for similar building components/items.

3.3.5.2 Where it is not possible to derive the method of measurement from the tabulated rules, the rules chosen may be bespoke. In such cases, the rule or rules derived shall be reiterated in full in either the preliminaries or in the bill of quantities (above the building components/items or items to which the rule relates).

3.3.6   Procedure where work cannot be quantified:

For the rules relating to work which cannot be quantified, refer to paragraph 2.9 (Non-measurable works) at Part 2 of these rules.

3.3.7   Procedure where exact type of product or component is not specified:

3.3.7.1 Where the exact type of product or component cannot be specified, an estimated price for the product or component shall be given in the description as a prime cost price (PC price). State, for example, 'Allow the PC price of £x per thousand delivered to site', 'Allow the PC price of £y per m^2 delivered to site', or 'Allow £z each delivered to site'.

3.3.7.2 Unless specifically stated otherwise in the bill of quantities or in these rules, the contractor shall be deemed to have allowed for all items listed in paragraph 3.3.3.13 in his or her priced rate for each building component/item incorporating a PC price.

3.3.7.3 PC prices shall exclude any allowance for the main contractor's overheads and profit, which are dealt with separately.

3.3.8   Procedure where quantity of work cannot be accurately determined:

3.3.8.1 Where work can be described and given in items in accordance with the tabulated rules of measurement but the quantity of work cannot be accurately determined, an estimate of the quantity shall be given and identified as a 'provisional quantity'.

3.3.8.2 Work items identified as a 'provisional quantity' shall be subject to remeasurement when they have been completed. The 'approximate quantity' shall be substituted by the 'firm quantity' measured, and the total price for that item adjusted to reflect the change in quantity. Where the variance between the 'provisional quantity' and the 'firm quantity' measured is less than 20 per cent, the rate tendered by the contractor shall not be subject to review. Where the variance is significant (i.e. 20 per cent or more), the rate can be reviewed to ensure that the rate is fair and reasonable to both the employer and contractor.

Tabulated work sections

1 Preliminaries (main contract)

Part A: Information and requirements

1.1 Project particulars

Sub-heading 1	Sub-heading 2	Information requirements	Supplementary information/notes
1 Project particulars	1 Name of project.	Short project title to be stated.	
	2 Nature of project.	Short description to be stated.	
	3 Location of project.	Full postal address to be stated.	
	4 Length of contract.	Period, in weeks, to be stated.	Where to be stated by the contractor, insert 'To be confirmed'.
	5 Names, addresses and points of contact of employer and consultants.	Function (e.g. architect), name of organisation, address, point of contact, telephone number and email address to be stated To include: (1) Employer. (2) Project sponsor (e.g. employer's internal project manager). (3) Project manager (if applicable). (4) Principal contractor (under CDM Regulations). (5) Person empowered by the contract to act on behalf of the employer (person's title to be given (e.g. contract administrator)). (6) CDM co-ordinator. (7) Quantity surveyor. (8) Consultants (separately identified). (9) Clerk of works (if required by the employer).	

1.2 Drawings

Sub-heading 1	Sub-heading 2	Information requirements	Supplementary information/notes
1 Drawings	1 List of drawings from which the bill of quantities was prepared.	Drawing number, including revision status, drawing title, and author to be stated.	Exceptions to be stated.
2 Other documents	1 Pre-construction information.	Explain how pre-construction information is dealt with (i.e. within the preliminaries bill or as a separate document). Cross-reference to pre-construction information document if separate document.	
	2 List of drawings and other documents relating to the contract but not included in the tender documents.	(1) Provide a list of drawings and other documents relating to the contract but not included in the tender documents, which may be seen by the contractor during the tender period. (2) Document title, reference, revision, date of issue and author to be stated. (3) Details of where documents can be seen to be stated.	

1.3 The site and existing buildings

Sub-heading 1	Sub-heading 2	Information requirements	Supplementary information/notes
1 The site		(1) Description of the site. (2) Reference to drawing, or drawings, showing the site boundaries and contractor's working area(s).	
2 Existing buildings on or adjacent to the site		Description of existing buildings on or adjacent to the site.	
3 Surrounding land/building uses	1 Address.	(1) Sub-heading 2 to identify land or building. (2) Use or activities carried out on the land or in building(s).	
4 Existing mains services	1 On the site.	List drawings and other applicable information.	
	2 Adjacent to the site.		
5 Soils and ground water		State information provided and where included in documentation (e.g. 'Annex B of the BQ' or 'as a separate document'). Cross-reference as necessary.	
6 Site investigation			

Sub-heading 1	Sub-heading 2	Information requirements	Supplementary information/notes
7 Health and safety file	1 Health and safety file.	(1) Availability for inspection.	
	2 Other documents.	(2) Arrangements for inspection.	
8 Health and safety hazards		Details of hazards that are or may be present on the site.	Where information provided in the pre-construction information, cross-reference accordingly.
9 Access to the site		(1) Description.	
		(2) Limitations.	
10 Parking		Details of employer's requirements in respect of parking and payments of fees and charges in connection of parking, including parking bay and parking meter suspensions.	
11 Use of the site		Details of any limitations.	
12 Site visits		(1) Purpose.	Include where information not given in the 'conditions of tender'.
		(2) Arrangements for site visit.	

1.4 Description of the Works

Sub-heading 1	Sub-heading 2	Information requirements	Supplementary information/notes
1 The Works		Description of the Works.	
2 Preparatory work by others		Description of any work that will be carried out by others under a separate contract before the start of work on site for this contract.	
3 Work by others concurrent with the contract		Description of the Works.	
4 Completion work by others			

1.5   The contract conditions

Sub-heading 1	Sub-heading 2	Information requirements	Supplementary information/notes
1 Conditions of contract	1 Form of contract title to be stated.	(1) Full title of the standard or bespoke form of contract, including edition, revision, and standard amendments applicable. (2) Schedule of clause/condition headings in the standard form of contract (see note (1)). (3) Reference to any amendments to clauses/conditions to standard form of contract (see note (1)). (4) Reference to any supplementary or special clauses/conditions to standard form of contract. (5) Insertions relating to articles of agreement, articles, recitals and contract particulars or abstract of particulars (see note (1)). (6) Employer's insurance responsibility. (7) Employer's requirements in respect of performance bonds. (8) Employer's requirements in respect of parent company guarantees. (9) Employer's requirements in respect of collateral warranties.	(1) Neither a schedule of clause/condition headings in the standard form of contract nor reference to any supplementary or special clauses/conditions to standard form of contract or details of insertions relating to articles of agreement, articles, recitals and contract particulars or abstract of particulars is required where addressed through a schedule of amendments to the standard conditions of contract. Notwithstanding this, reference to the schedule of amendments is to be made in this section of the preliminaries bill. (2) Where bespoke, or uncommon, forms of contract are used – a copy is to be appended to the bill of quantities or included as part of the tender.

1.6   Employer's requirements: Provision, content and use of documents

Sub-heading 1	Sub-heading 2	Information requirements	Supplementary information/notes
1 Definitions and interpretations	1 Definitions.	Explain how to interpret key words, terms, phrases and synonyms used in the preliminaries and specification.	
	2 Communication.	Definition and format of communications, and timing of response.	

Sub-heading 1	Sub-heading 2	Information requirements	Supplementary information/notes
	3 Products.	Definition and/or meaning.	
	4 Site equipment.		
	5 Drawings.		
	6 Contractor's choice.		
	7 Contractor's designed work.		
	8 Submit proposals.		
	9 Terms used in specification.	Definitions and/or meaning of key words, terms, phrases and synonyms used in the specification.	
	10 Manufacturer and product references.	(1) Definition of terms. (2) Version of manufacturer's technical literature applicable to tender and contract (e.g. current on the date of invitation to tender).	
	11 Substitution of products.	(1) Definition and/or meaning of substitute/alternative products. (2) Process for acceptance and rejection of substitute/alternative products.	
	12 Cross-references.	Explain method of cross referencing used.	
	13 Referenced documents.	Order of precedence of referenced documents.	
	14 Equivalent products.	Definition and/or meaning.	
	15 Substitution of standards.	(1) Definition and/or meaning. (2) Process for acceptance and rejection of substitute standards.	
	16 Currency of documents.	Version of published documents, including revisions and amendments, applicable to tender and contract (e.g. current on the date of invitation to tender).	
	17 Product sizes.	(1) General definition of product sizes. (2) Exceptions to general definition.	Products to be specified by their co-ordinating size. Exceptions to this rule to be stated.

Sub-heading 1	Sub-heading 2	Information requirements	Supplementary information/notes
2 Documents provided on behalf of employer	1 Additional copies of drawings and documents.	Describe procedure.	
	2 Dimensions.	Explain ownership of scaled dimensions	
	3 Measured quantities.	Explain precedence of measured quantities.	
	4 The specification.	(1) Reference the specification, or specifications, the preliminaries. (2) Explain the method used to cross reference specification clauses on or in other tender/contract documents.	
	5 Divergence from the statutory requirements.	Method for dealing with divergence from the statutory requirements should they occur.	
	6 Employer's policy documents.	Requirements in respect of compliance with the employer's policies. Examples include: – environmental. – sustainability. – corporate social responsibilities (CSR). – health and safety.	
3 Documents provided by the contractor, subcontractors and suppliers	1 Design information.	(1) General requirements, including design management and programming requirements. (2) Specific requirements in respect of design documents and information. (3) Format. (4) Number of copies. (5) Submission requirements.	
	2 Production information.	(1) General requirements in respect of production information. (2) Format. (3) Number of copies. (4) Submission requirements.	
	3 As-built/as-installed drawings and information.	(1) General requirements. (2) Submission requirements. (3) Number of copies.	
	4 Technical literature.	(1) Literature to be maintained. (2) Requirements for literature to be available on site.	

Sub-heading 1	Sub-heading 2	Information requirements	Supplementary information/notes
	5 Maintenance instructions and guarantees.	(1) Information requirements. (2) Format. (3) Number of copies. (4) Submission requirements. (5) Storage and information management. (6) Requirements in respect of emergency and/or out of normal working call-out services, including requirements for contact details and extent of cover.	Requirements relating to 'management information systems' to be stated under storage and information management requirements.
	6 Energy rating calculations.	(1) Information requirements. (2) Number of copies. (3) Submission requirements.	
	7 Environmental assessment information.	(1) Scheme type. (2) Environmental targets (in respect of site activities and the works). (3) Information requirements. (4) Format. (5) Number of copies. (6) Submission requirements.	
	8 Documents required before Practical Completion.	Specific requirements.	
4 Document and data interchange	1 Electronic data interchange (EDI).	(1) Types and classes of communication. (2) Parties communication between. (3) Requirements.	

1.7 Employer's requirements: Management of the works

Sub-heading 1	Sub-heading 2	Information requirements	Supplementary information/notes
1 Employer's requirements – generally	1 Supervision.	Specific requirements.	
	2 Considerate constructors scheme.	(1) Registration requirements. (2) Contact details. (3) Compliance requirements.	
	3 Insurance.	(1) Documentary evidence required. (2) When required.	
	4 Insurance claims.	(1) Requirements for notifying events. (2) Requirements for indemnifying the employer.	
	5 Climatic conditions.	Records to be maintained by contractor.	
	6 Ownership of materials/products arising from the works.	Requirements in respect of ownership, and removal from site.	Where to become the property of the contractor; consider seeking credit for materials/products arising from the works.
2 Programme/progress	1 Programme.	(1) Format and content. (2) Exclusions. (3) Document control requirements. (4) Submission requirements.	
	2 Revised programme.	Specific requirements for re-profiling and reissuing programme.	
	4 Commencement of work.	Notice period to be given before the commencement of work on site.	
	5 Monitoring progress.	(1) Employer's specific requirements in respect of reporting and avoiding potential delay. (2) List and description of key performance indicators (KPIs) to be maintained by contractor. (3) Requirements for reporting against KPIs. (4) Actions of contractor if KPIs not achieved.	
	6 Notification of compensation event(s).	Employer's specific requirements in respect of contractor notifying events that compensate potential delay.	

Sub-heading 1	Sub-heading 2	Information requirements	Supplementary information/notes
	7 Project progress meetings.	(1) General requirements. (2) Proposed agenda. (3) Frequency. (4) Location. (5) Accommodation availability. (6) Attendees. (7) Chairperson.	
	8 Contractor's progress report.	(1) Form and content of report. (2) Method of presentation. (3) Submission requirements.	
	9 Contractor's site meetings.	General requirements.	
	10 Photographic records.	(1) Image format. (2) Frequency. (3) Number of locations. (4) Number of images from each location. (5) Other requirements.	
	11 Early possession.	Employer's specific requirements for early possession and taking over parts of the works before practical completion.	
	12 Notice of completion.	(1) General requirements. (2) Minimum period of notice to be given by contractor to be stated.	
	13 Extensions of time.	(1) Requirements in respect of notification by the contractor. (2) Submission requirements.	
3 Cost control	1 Cash flow forecast.	(1) Basis of cash flow forecast. (2) Frequency. (3) Submission requirements.	
	2 Removal/replacement of existing work.	(1) Location. (2) Extent. (3) Requirements in respect of execution.	

Sub-heading 1	Sub-heading 2	Information requirements	Supplementary information/notes
	3 Proposed instructions.	(1) Requirements in respect of estimates. (2) Content of estimates. (3) Other requirements.	
	4 Measurement of covered work.	(1) General requirements. (2) Notice period to be given by contractor before covering works which are to be measured.	
	5 Daywork vouchers.	(1) Notice period to be given by contractor before commencement of work to be carried out on a daywork basis. (2) Submission requirements.	
	6 Interim applications, valuations and payments.	(1) Process for agreeing interim payments and payment dates. (2) Content and submission requirements for contractor's interim applications for payment. (3) Employer's specific invoicing requirements.	
	7 Payment for products not incorporated into the works.	Information/evidence of freedom from title required from contractor in respect of products stored on-site before payment is to be considered by employer.	
	8 Payment for products stored off-site.	Information/evidence of freedom from title required from contractor in respect of products stored off-site before payment is to be considered by employer.	
	9 Labour and equipment returns.	(1) Records to be maintained by the contractor. (2) Content of records. (3) Submission of records.	

1.8   Employer's requirements: Quality standards and quality control

Sub-heading 1	Sub-heading 2	Information requirements	Supplementary information/notes
1 Standards of products and executions	1 Incomplete information.	(1) Requirements for dealing with products/materials are not fully specified. (2) Status of works where the extent is not fully documented. (3) Status of omissions or errors in description and/or quantity.	
	2 Workmanship skills.	Requirements in respect of: (1) Appropriateness of contractor's operatives. (2) Registration schemes to which contractor's operatives should belong. (3) Evidence of scheme registration requirements. (4) Other requirements.	
	3 Quality of products.	Requirements in respect of: (1) Using new and recycled products. (2) Supply of products. (3) Tolerances. (4) Deterioration prevention. (5) Other requirements.	
	4 Quality of execution.	Requirements in respect of: (1) Fixing, application and installation of products, including alignment. (2) Colour batching. (3) Dimensions. (4) Finished work. (5) Location and fixing of products. (6) Other requirements.	
	5 Compliance.	Requirements in respect of: (1) Compliance with proprietary specifications. (2) Compliance with performance specifications. (3) Other requirements.	

Sub-heading 1	Sub-heading 2	Information requirements	Supplementary information/notes
	6 Inspections.	Requirements for inspecting products and work executed.	
	7 Related work.	Requirements in respect of co-ordinating workmanship of trades.	
	8 Manufacturer's recommendations/instructions.	(1) Requirements in respect of compliance. (2) Version of manufacturer's recommendations/instructions applicable to tender and contract (e.g. current on the date of invitation to tender).	
	9 Water for the works.	General requirements.	
2 Samples/Approvals	1 Samples.	Requirements in respect of samples of products, work executed and mock-ups.	
	2 Approval of product samples.	(1) General requirements (including programming requirements) for submission of, inspection of, and tests on samples. (2) Definition of approval in context of samples. (3) Retention of complying samples, including storage requirements. (4) Other requirements.	
	3 Approval of work executed/mock-ups.	(1) General requirements (including programming requirements) for submission of, inspection of, tests on, work executed and mock-ups. (2) Definition of approval in context of work executed and mock-ups. (3) Retention of complying work executed and mock-ups, including storage requirements. (4) Other requirements.	

Sub-heading 1	Sub-heading 2	Information requirements	Supplementary information/notes
3 Accuracy/Setting out	1 Accuracy of instruments.	General requirements.	
	2 Setting out.		
	3 Appearance and fit.	Requirements in respect of tolerances and dimensions.	
	4 Critical dimensions.	Details of critical dimensions.	
	5 Levels of structural floors.	Maximum tolerances for designed levels to be stated.	
	6 Record drawings.	(1) Requirements in respect of recording details of grid lines, setting-out stations, benchmarks and profiles. (2) Information retention requirements. (3) Submission requirements. (4) Other requirements.	
4 Services	1 Services regulations.	General requirements.	
	2 Water regulations/byelaws notification.		
	3 Water regulations/byelaws contractor's certificate.	(1) Content of certificate. (2) Submission requirements. (3) Other requirements.	
	4 Electrical installation certificate.	(1) Submission requirements. (2) Other requirements.	
	5 Gas, oil and solid fuel appliance installation certificate.	(1) Content of certificate. (2) Submission requirements. (3) Other requirements.	
	6 Mechanical and electrical services.	(1) Requirements in respect of final tests and commissioning. (2) Requirements in respect of Building Regulations notice. (3) Other requirements.	
5 Supervision/inspection/defective work	1 Supervision.	(1) General requirements. (2) Notice period for replacement of contractor's person in charge by contractor.	
	2 Co-ordination of mechanical and electrical engineering services.	(1) General requirements. (2) Requirements for documentary evidence of contractor's staff.	

Sub-heading 1	Sub-heading 2	Information requirements	Supplementary information/notes
	3 Overtime working.	General requirements.	
	4 Defects in existing work.	(1) Process for dealing with undocumented defects. (2) Process for dealing with documented remedial work.	
	5 Access for inspection.	General requirements.	
	6 Tests and inspections.	(1) Timing requirements. (2) Records required. (3) Submission requirements. (4) Other requirements.	
	7 Air permeability.	(1) Method. (2) Performance requirements. (3) Submission requirements (results). (4) Other requirements.	
	8 Continuity of thermal insulation.	(1) General requirements. (2) Content of reports. (3) Submission requirements. (4) Number of copies. (5) Other requirements.	
	9 Resistance to passage of sound.	(1) Method. (2) Compliance requirements. (3) Submission requirements. (4) Other requirements.	
	10 Energy performance certificate.	(1) General requirements. (2) Format of certificate and report. (3) Submission requirements. (4) Other requirements.	
	11 Proposals for rectification of defective products/executions.	General requirements.	
	12 Measures to establish acceptability.		

Sub-heading 1	Sub-heading 2	Information requirements	Supplementary information/notes
	13 Quality control.	(1) Procedural requirements. (2) Records required. (3) Content of records. (4) Other requirements.	
6 Work at or after completion	1 Work before completion.	(1) General requirements. (2) Cleaning requirements, including cleaning materials and methods of cleaning. (3) Requirements for rectifying minor faults. (4) Requirements in respect of moving parts. (5) Other requirements.	
	2 Security at completion.	(1) General requirements. (2) Requirements in respect of keys.	
	3 Making good/rectification of defects.	(1) Access arrangements. (2) Notice periods. (3) Completion requirements.	
	4 Highway/sewer adoption.	(1) Description of work to be adopted. (2) Requirements in respect of work for adoption.	

1.9 Employer's requirements: Security, safety and protection

Sub-heading 1	Sub-heading 2	Information requirements	Supplementary information/notes
1 Security/health and safety	1 Pre-construction information.	Location of information.	
	2 Execution hazards.	(1) Management requirements for common hazards. (2) Details of significant hazards incorporated in the design of the project.	
	3 Product hazards.	(1) Requirements in respect of hazardous substances. (2) Management requirements for common hazards. (3) Details of significant hazards in specified construction materials.	

Sub-heading 1	Sub-heading 2	Information requirements	Supplementary information/notes
	4 Construction phase health and safety plan.	(1) General requirements. (2) Content. (3) Submission requirements.	
	5 Security.	(1) Requirements for protecting the site, the works, products, materials, and existing buildings affected by the works, from damage and theft. (2) Requirements for preventing unauthorised access to the site, the works, and adjoining property. (3) Description of any special security requirements.	
	6 Stability.	Requirements for maintaining the stability and structural integrity of the works and adjoining property during the contract.	
	7 Occupied premises.	(1) Extent existing buildings will be occupied and/or used during the contract. (2) Method of working. (3) Requirements in respect of overtime working.	
	8 Passes.	(1) Details of controlled areas. (2) Procedures for obtaining and returning passes.	
	9 Occupier's rules and regulations.	(1) General requirements. (2) Location of rules and regulations. (3) Arrangements for inspection.	
	10 Use of mobile telephones.	Requirements relating to use of mobile telephones on-site.	
	11 Employer's representatives site visits.	(1) Safety requirements. (2) Protective clothing and/or equipment requirements.	
	12 Working precautions/restrictions.	(1) Details of hazardous areas. (2) Permit to work requirements.	
2 Protection against	1 Explosives.	(1) Use. (2) Details of restrictions.	

Sub-heading 1	Sub-heading 2	Information requirements	Supplementary information/notes
	2 Noise consent by local authority.	(1) General requirements.	
	3 Noise control.	(2) Special requirements.	
	4 Pollution control.		
	5 Pesticides.	(1) Use.	
		(2) Details of restrictions.	
		(3) Disposal requirements.	
		(4) Operatives competency requirements.	
	6 Nuisance.	(1) General requirements.	
	7 Asbestos containing materials (ACM)s.	(2) Special requirements.	
	8 Antiquities.		
	9 Fire prevention.	(1) General requirements.	
		(2) Details of standards to which contractor is to comply.	
	10 Smoking on-site.	(1) General requirements.	
	11 Burning on-site.	(2) Special requirements.	
	12 Moisture.	(1) Requirements for preventing wetness and dampness.	
		(2) Requirements for drying out.	
	13 Infected timber/contaminated materials.	General requirements.	
	14 Waste.	(1) Definition of waste.	
		(2) General requirements, including minimizing waste, removing waste and excluding waste from voids and cavities in the construction.	
		(3) Requirements in respect of 'site waste management plans' (SWMP).	
		(4) Documentation requirements.	
		(5) Details of key performance indicator (KPI) data to be provided.	
		(6) Information and documentation submission requirements.	

Sub-heading 1	Sub-heading 2	Information requirements	Supplementary information/notes
	15 Electromagnetic interference.	General requirements.	
	16 Laser equipment.		
	17 Power actuated fixing systems.		
	18 Invasive species.	(1) General requirements for the prevention of invasive species (e.g. plants and animals). (2) Details of any special precautions required. (3) Requirements in respect of discovery and reporting.	
3 Protection	1 Existing services.	(1) Requirements in respect of notifying services authorities, statutory undertakers and/or adjoining adjacent owners. (2) Requirements for identification of existing services. (3) Requirements should damage occur to existing services. (4) Other requirements.	
	2 Roads and footpaths.	(1) General requirements. (2) Requirements should damage occur to existing roads and/or footpaths. (3) Other requirements.	
	3 Existing topsoil and subsoil.	(1) Requirements for preventing over compaction of existing topsoil and subsoil. (2) Details of protective measures.	
	4 Retained trees, shrubs and grassed areas.	(1) Details of protective measures. (2) Requirements should damage occur to retained trees, shrubs and grassed areas. (3) Other requirements.	
	5 Areas of retained trees.	Details of protective measures.	
	6 Wildlife species and habitats.	(1) General requirements. (2) Details of protective measures. (3) Other requirements.	
	7 Existing features.	(1) Details of protective measures. (2) Special requirements.	

Sub-heading 1	Sub-heading 2	Information requirements	Supplementary information/notes
	8 Existing work.	(1) Details of protective measures. (2) Requirements for removing and replacing existing work.	
	9 Building interiors.	Details of protective measures.	
	10 Existing furniture, fittings and equipment.	(1) Details of protective measures. (2) Extent of removal work to be carried out by the employer.	
	11 Especially valuable and vulnerable items.	(1) Details of protective measures. (2) Extent of removal work to be carried out by the employer.	
	12 Adjoining property.	Permission requirements.	
	13 Adjoining property restrictions.	(1) Precautions to be taken by the contractor. (2) Consequence of damage.	
	14 Existing structures.	General requirements.	
	15 Materials for recycling and/or reuse.	(1) Sorting and damage prevention requirements. (2) Storage requirements.	

1.10   Employer's requirements: Specific limitations on method, sequence and timing

Sub-heading 1	Sub-heading 2	Information requirements	Supplementary information/notes
1 General		Statement explaining that limitations described in this section of the preliminaries bill are supplementary to limitations described or implicit in information given in other sections or on the drawings.	
2 Design constraints		Details of any design constraints.	
3 Method/sequence of work		Specific limitations relating to method and sequence of working, including phasing requirements to be included in the programme.	
4 Use or disposal of materials found		Specific limitations.	

Sub-heading 1	Sub-heading 2	Information requirements	Supplementary information/notes
5 Use or disposal of materials found		Specific limitations.	
6 Working hours		(1) Definition of work hours. (2) Definition of normal working hours.	
7 Completion of any section or part of the works	1 Employer requirements for possession.	(1) General requirements. (2) Special requirements.	
	2 Remainder of the works.	Requirements in respect of: (1) Provision of services. (2) Fire precautions. (3) Means of escape and safe access. (4) Other requirements.	

1.11 Employer's requirements: Site accommodation/services/facilities/temporary work

Sub-heading 1	Sub-heading 2	Information requirements	Supplementary information/notes
1 Generally	1 Spoil heaps, temporary works and services.	Requirements in respect of the siting of spoil heaps and the maintenance, alteration, movement and removal of temporary works.	
2 Site accommodation	1 Room for meetings.	(1) Specific requirements, including furniture and equipment, to be stated. (2) Cleaning and maintenance requirements.	
	2 Site offices.		
	3 Off-site offices/room for meetings.	(1) Specific requirements, including furniture and equipment. (2) Preparatory works, including painting, decoration and applied finishings (e.g. carpet). (3) Cleaning and maintenance requirements.	
	4 Sanitary accommodation.	Specific requirements, including furniture and equipment.	

Sub-heading 1	Sub-heading 2	Information requirements	Supplementary information/notes
	5 Use of accommodation/land not included in the site.	(1) Identification of accommodation and or land that may be used by the contractor for the duration of the contract without charge. (2) Limitations/restrictions on use. (3) Requirements for temporary adaptations. (4) Cleaning and maintenance requirements. (5) Accommodation/land use. (6) Location of accommodation/land. (7) Reinstatement requirements on vacation of accommodation/land.	
	6 Car parking.	Specific requirements for the provision of car parking for representatives of the employer.	
3 Services and facilities	1 Lighting.	Specific requirements for the provision temporary lighting for finishing work and inspection.	
	2 Lighting and power.	(1) Requirements in respect of: (a) Use of employer's mains supply. (b) Responsibility for continuity of supply. (2) Metering requirements. (3) Location of supply point. (4) Available capacity, frequency, phase and current type. (5) Conditions/restrictions imposed on the contractor.	
	3 Water.	(1) Requirements to be stated in respect of (a) Use of employer's mains supply. (b) Responsibility for continuity of supply. (2) Metering requirements. (3) Source of supply. (4) Location of supply point. (5) Conditions/restrictions imposed on the contractor.	
	4 Contractor's on-site telephones.	(1) Date to be installed. (2) Responsibility for paying installation and all rental charges, including paying the cost of calls. (3) Requirements for disseminating telephone number. (4) Other requirements.	

Sub-heading 1	Sub-heading 2	Information requirements	Supplementary information/notes
	5 Mobile telephones.	Requirements in respect of: (1) Contractor's staff required to be provided with mobile telephones. (2) Responsibility for paying installation and all rental charges, including paying the cost of calls. (3) Requirement for disseminating telephone numbers. (4) Other requirements.	
	6 Telephones.	(1) System requirements. (2) Date to be installed. (3) Responsibility for paying installation and all rental charges, including paying the cost of calls. (4) Other requirements.	
	7 Fax installation.		
	8 Computers.	(1) System requirements, including computers, software, printers, cables and consumables. (2) Responsibility for paying the cost of consumables. (3) Date to be installed. (4) Other requirements.	
	9 Email and internet facilities.	(1) System requirements. (2) Date to be installed. (3) Responsibility for paying installation and all rental charges. (4) Other requirements.	
	10 Photocopier.	Employer's requirements.	
	11 Temperature and humidity.	Levels to be maintained by the contractor.	

Sub-heading 1	Sub-heading 2	Information requirements	Supplementary information/notes
	12 Use of permanent heating systems.	(1) Confirmation or otherwise that the contractor is permitted to use permanent heating systems for drying out the works, services and controlling temperature and humidity levels. (2) Requirements for operation, maintenance and remedial work. (3) Requirements for contractor to arrange supervision of use by subcontractor and indemnification of subcontractor. (4) Other requirements.	
	13 Beneficial use of permanent installed systems.	Details of services systems, including lifts and sanitary installations, which can be used by the contractor to complete the works.	
	14 Meter readings.	Requirements for obtaining meter readings.	
	15 Thermometers.	Requirements for providing maximum and minimum thermometers for measuring atmospheric shade temperature.	
	16 Surveying equipment.	Requirements for providing surveying equipment.	
	17 Personal protective equipment.	Specific requirements for those acting on behalf of the employer.	
	18 Other requirements.	Specific requirements.	
Temporary works	1 Roads, hard standings and footpaths.	Requirements in respect of permanent roads, hard standings and footpaths on the site, including restrictions on use and protective and/or remedial measures.	
	2 Temporary works.	Specific requirements (e.g. fences, hoardings, screens and roofs).	

Sub-heading 1	Sub-heading 2	Information requirements	Supplementary information/notes
	3 Temporary protection measures to existing trees/vegetation.	(1) Location of temporary protection (by reference to drawing) (2) Standards of protective barriers and any other applicable physical protection measures. (3) Design details of physical protection measures (by reference to drawing). (4) Areas of structural landscaping to be protected from construction operations. (5) Requirements for maintaining integrity of protection for the duration of the works. (6) Requirements for removing protection on completion of the works.	
	4 Name boards.	Specific requirements.	
	5 Advertising.		
	6 Other requirements.		

1.12 Employer's requirements: Operation/maintenance of finished building

Sub-heading 1	Sub-heading 2	Information requirements	Supplementary information/notes
1 Operation and maintenance manual	1 Generally.	(1) Purpose. (2) Scope. (3) Responsibility for preparation. (4) Information to be provided by others. (5) Review process. (6) Number of copies. (7) Latest date for submission of final manual. (8) As-built/as-installed drawings: (a) format and standard (b) number of copies.	The operation and maintenance information, the health and safety file, and all other information can be combined as a single document. In this case, the document can be referred to as the 'building manual'.
	2 Content.	Specific requirements.	

Sub-heading 1	Sub-heading 2	Information requirements	Supplementary information/notes
2 Health and safety file	1 Generally.	(1) Purpose. (2) Scope. (3) Responsibility for preparation. (4) Information to be provided by the contractor, where not responsible for preparation. (5) Review process. (6) Number of copies. (7) Latest date for submission of final file.	
	2 Content.	Specific requirements.	
3 Web-based information management system		Specific requirements.	
4 Presentation of documents	1 Operation and maintenance manual.	Specific requirements in respect of format and presentation.	
	2 Health and safety file.		
	3 Other documents.	Specific requirements.	
5 Other employer specific requirements	1 Maintenance services.	Specific requirements for post completion maintenance (planned and reactive).	
	2 Information for commissioning services.	Specific requirements.	
	3 Training.		
	4 Spare parts.		
	5 Tools.		
6 Other information		Specific requirements.	

1 Preliminaries (main contract)

Part B: Pricing schedule

1.1 Employer's requirements

1.1.1 Site accommodation

Component	Included/notes on pricing	Unit	Pricing method	Excluded
1 Site accommodation	Site accommodation for the employer and the employer's representatives where separate from main contractor's site accommodation, including: – site offices. – sanitary accommodation. – welfare facilities. – foundations to site accommodation. – temporary drainage to accommodation. – temporary services. – intruder alarms. Type and extent of accommodation to be provided to be stated; with each type separately quantified.			Site accommodation, furniture and equipment, telecommunication and IT systems for the employer and the employer representatives where an integral part of the main contractor's site accommodation (included in section 1.2: in contractor's cost items, as appropriate).
	1 Bringing to site and installing, including all temporary drainage, services and intruder alarms.	item	Fixed charge.	
	2 Adaptations/alterations during works.			
	3 Dismantling and removing from site, including rectifying any damage.			
	4 Maintaining.	weeks	Time-related charge.	
	5 Cleaning.			
	6 Charges.			
	7 Off-site rented temporary accommodation.			
	8 Rectifying damage to off-site rented temporary accommodation.	item	Fixed charge.	

Component	Included/notes on pricing	Unit	Pricing method	Excluded
2 Furniture and equipment	Furniture and equipment for the employer and the employer's representatives where separate from main contractor's site accommodation. For example, desks, chairs, meeting table and chairs, cupboards, kettles, coffee maker, photocopier and consumables.			
	1 Bringing to site and installing.	item	Fixed charge.	
	2 Cleaning.	week	Time-related charge.	
	3 Charges.			
	4 Dismantling and removing from site.	item	Fixed charge.	
3 Telecommunications and IT systems	Telecommunication and IT systems for the employer and the employer's representatives where separate from main contractor's site accommodation, including telephones, fax machines, photocopier, computers, printers and consumables.			
	1 Purchase charges.	nr	Fixed charge.	
	2 Hire charges.	week	Time-related charge.	
	3 Consumables.			

1.1.2 Site records

Component	Included	Unit	Pricing method	Excluded
1 Site records	1 Operation and maintenance manuals.	item	Fixed charge.	
	2 Compilation of health and safety file.			
2 Web-based information management system	1 Provision of system.			
	2 Uploading of data.			
	3 Training of building user's staff in the operation of the web-based management system.			

1.1.3 Completion and post-completion requirements

Component	Included	Unit	Pricing method	Excluded
1 Handover requirements	1 Training of building user's staff in the operation and maintenance of the building engineering services systems. 2 Provision of spare parts for maintenance of building engineering services. 3 Provision of tools and portable indicating instruments for the operation and maintenance of building engineering services systems.	item	Fixed charge.	
2 Operation and maintenance services	1 Operation and maintenance of building engineering services installations, mechanical plant and equipment and the like during the defects liability period, period for rectifying defects, maintenance period or other specified period (i.e. additional services that are normally required by the contract).	week	Time-related charge.	
3 Landscape management services	1 Maintenance of internal and external planting.			

1.2 Main Contractor's cost items

1.2.1 Management and staff

Component	Included	Unit	Pricing method	Excluded
1 Project-specific management and staff	Main contractor's project specific management and staff such as: 1 Project manager/director. 2 Construction manager. 3 Supervisors, including works/trade package managers, building services engineering managers/co-ordinators and off-site production managers. 4 Health and safety manager/officers. 5 Commissioning manager (building engineering services).	week (number of staff by number of man hours per week by number of weeks)	Time-related charge.	1 Security staff (included in section 1.2.4: Security).

Component	Included	Unit	Pricing method	Excluded
	6 Planning/programming manager and staff.			
	7 Senior/managing quantity surveyor.			
	8 Project/package quantity surveyors.			
	9 Procurement manager.			
	10 Design manager.			
	11 Project engineers.			
	12 Environmental manager.			
	13 Temporary works design engineers.			
	14 Materials management staff (e.g. storeman).			
	15 Administrative staff, including secretary, document controllers, finance clerks and the like.			
	16 Other management and staff.			
2 Visiting management and staff	1 Managing director, regional director, operations director, commercial director and the like.			1 Visiting management and staff for which an allowance has been made within the main contractor's overheads.
	2 Quality manager.			
	3 Contracts/commercial manager.			
	4 Health and safety manager.			
	5 Environmental manager/consultant.			
	6 Other visiting management and staff.			
3 Extraordinary support costs	1 Legal advice costs (i.e. solicitors).	item	Fixed charge.	1 Extraordinary support costs for which an allowance has been made within the main contractor's overheads.
	2 Recruitment costs.			
	3 Team building costs.			
	4 Other extraordinary support costs.			
	5 Day transport.	week (number of days per week by number of weeks)	Time-related charge.	
	6 Personnel transport (i.e. transportation of work operatives to site).			

Component	Included	Unit	Pricing method	Excluded
	7 Temporary living accommodation (e.g. long/medium term accommodation costs). 8 Subsistence payments. 9 Out of normal hours working, including non-productive overtime allowances.	week (number of staff by number of days per week by number of weeks)		
4 Staff travel	Costs associated with off-site visits such as: 1 Visits to employer's and consultants' offices.	nr (number of occasions)	Fixed charge.	
	2 Visits to subcontractors' offices/works.			
	3 Overseas visits.			
	4 Accommodation charges and overnight expenses.			

1.2.2 Site establishment

Component	Included	Unit	Pricing method	Excluded
1 Site accommodation	Main contractor's and common user temporary site accommodation such as: – offices. – conference/meeting rooms. – canteens and kitchens. – drying rooms. – toilets and washrooms. – first aid room. – laboratories. – workshops. – secure stores. – compounds, including containers for material storage. – security control room. – stairs and office staging. Type and extent of accommodation to be provided to be stated, with each type separately quantified.			1 Employer's accommodation, where not an integral part of the main contractor's site accommodation (included in section 1.1.1: Site accommodation). 2 Temporary bases, foundations and provision of drainage and services to temporary site accommodation (included in component 1.2.2.2: Temporary works in connection with site establishment). 3 Service provider's charges for temporary services (included in section 1.2.12: Fees and charges). 4 Rates for temporary services (included in section 1.2.12: Fees and charges).

Component	Included	Unit	Pricing method	Excluded
	1 Purchase charges.	item	Fixed charge.	
	2 Hire charges.	week	Time-related charge.	
	3 Employer's accommodation, where an integral part of the main contractor's site accommodation.			
	4 Delivery of temporary site accommodation to site, erection, construction and removal.	item	Fixed charge.	
	5 Temporary accommodation made available by the employer.	week	Time-related charge.	
	6 Intruder alarms.	item	Fixed charge.	
	7 Land/property rental where site accommodation located off-site.	week	Time-related charge.	
	8 Alterations and adaptations to site accommodation, including partitioning, doors, painting and decorating, and the like.	item	Fixed charge.	
	9 Relocation and alterations of temporary accommodation during construction stage.			
	10 Reinstating temporary site accommodation to original condition prior to removal from site.			
	11 Removal of site accommodation and temporary works in connection with site accommodation.			
2 Temporary works in connection with site establishment	1 Temporary bases and foundations for site accommodation, including maintenance and reinstatement of existing surfaces on completion of the works.	m²	1 Fixed charge. 2 Time-related charge.	1 Provision of temporary services to site establishment (included in section 1.2.3: Temporary services). 2 Provision of temporary drainage to site establishment (included in section 1.2.3: Temporary services). 3 Hoardings, fans, fencing and the like to site boundaries and to form site compounds (included in section 1.2.4: Security (Hoardings, fences and gates).
	2 Connections to temporary service, including maintenance and removal on completion of the works.	nr		
	3 Connections to temporary drainage, in including maintenance and removal on completion of the works.	nr		
	4 Temporary site roads, paths and pavings (including on-site car parking), including reinstatement of existing surfaces on completion of the works.	m		
	5 Temporary surface water drainage to temporary site roads, paths and pavements, including maintenance and removal on completion of the works.	m		

Component	Included	Unit	Pricing method	Excluded
3 Furniture and equipment	1 Workstations for staff, including maintenance.	nr	1 Fixed charge. 2 Time-related charge.	1 Telephone and fax installations (included in section 1.2.3: Temporary services). 2 Computers and IT associated equipment (included in component 1.2.2.4: IT systems).
	2 General office furniture, including maintenance.	item		
	3 Conference/meeting room furniture, including maintenance.			
	4 Photocopiers, including purchase/rental, maintenance and other running costs.			
	5 Canteen furniture, including maintenance.			
	6 Canteen equipment, including purchase/rental, maintenance and other running costs.			
	7 Floor coverings, including maintenance.			
	8 Water dispensers, including purchase/rental, maintenance and other running costs.			
	9 Heaters, including maintenance of heaters.			
	10 Other office equipment, including maintenance.			
	11 Removal of furniture and equipment.			
	12 Maintenance furniture and floor covering.			

Component	Included	Unit	Pricing method	Excluded
4 IT systems	1 Computer hardware, including purchase/rental, installation, initial set up, maintenance and running costs, such as: – desktop computers and laptop computers. – CAD stations. – server and network equipment. – printers and plotters. – other computer system hardware.	item	1 Fixed charge. 2 Time-related charge.	1 Computer and printer consumables (included in component 1.1.2.5: Consumables and services). 2 Document management, including electronic data management systems (EDMS) (included in component 1.1.2.6: Brought in services).
	2 Software and software licences.			
	3 Modem lines, modems and connections (i.e. email and internet capability).			
	4 WAN lines and connections (if on WAN).			
	5 Line rental charges.	week	Time-related charge.	
	6 Internet/website addresses.	nr	Fixed charge.	
	7 Internet service provider (ISP) charges.			
	8 Line calls charges.	week	Time-related charge.	
	9 IT support and maintenance.			
5 Consumables and services	1 Stationery.	week	Time-related charge.	
	2 Computer and printer consumables (e.g. ink cartridges).			
	3 Postage.			
	4 Courier charges.			
	5 Tea, coffee, water bottles and the like.			
	6 First aid consumables.			
	7 Photocopier consumables (e.g. paper and toners).			
	8 Fax consumables (e.g. paper and toners).			
	9 Drawing printer consumables (e.g. ink cartridges).			

Component	Included	Unit	Pricing method	Excluded
6 Brought-in services	Services outsourced by the main contractor such as: 1 Catering. 2 Equipment maintenance. 3 Document management, including management information systems and electronic data management systems (EDMS). 4 Printing (purchasing), including reports and drawings. 5 Staff transport. 6 Off-site parking charges. 7 Meeting room facilities. 8 Photographic services. 9 Other.	week	Time-related charge.	
7 Sundries	1 Main contractor's signboards. 2 Safety and information notice boards. 3 Fire points. 4 Shelters. 5 Tool stores. 6 Crane signage. 7 Employer's composite signboards.	item	Fixed charge.	

1.2.3 Temporary services

Component	Included	Unit	Pricing method	Excluded
1 Temporary water supply	1 Temporary connections. 2 Distribution equipment, installation and adaptations. 3 Meter charges.	nr item week	1 Fixed charge. 2 Time-related charge.	
2 Temporary gas supply	1 Gas connection. 2 Distribution equipment, installation and adaptations. 3 Charges. 4 Bottled gas.	nr item week	1 Fixed charge. 2 Time-related charge.	
3 Temporary electricity supply	1 Temporary connections. 2 Temporary electrical supply for tower cranes. 3 Charges – power consumption for site establishment. 4 Charges – power consumption for the works. 5 Distribution equipment, installation and adaptations. 6 Attendance. 7 Uninterrupted power supply (UPS). 8 Temporary substation modifications.	nr item item item nr (number of man hours per week by number of weeks) item	Fixed charge. Time-related charge. Fixed charge. Time-related charge. Fixed charge.	
4 Temporary telecommunication systems	1 Landlines (including connection and rental charges), including: – telephone and fax lines – ISDN lines. 2 Telephone and facsimile equipment (including connection and rental charges), including: – PABX equipment. – handsets, including purchase or rental. – fax machines, including purchase or rental. – installation of equipment. – maintenance of equipment.	item	1 Fixed charge. 2 Time-related charge.	1 Fax consumables (included in component 1.1.2.5: Consumables and services).

Component	Included	Unit	Pricing method	Excluded
	3 Mobile (cellular) phones, including: – mobile phones, including purchase or rental and connection charges. – spare batteries. – mobile phone charges.			
	4 Telephone charges, including: – telephone call charges. – fax charges. – fax and telephone consumables.			
	5 Radios (including purchase or rental charges), including: – base set. – handsets and chargers. – repairs and maintenance. – licences. – spare batteries.			
5 Temporary drainage	1 Temporary mains.	item	1 Fixed charge. 2 Time-related charge.	
	2 Septic tanks.	nr		
	3 On-site treatment plant.	item		
	4 Holding tanks.	nr		
	5 Sewage pumping.			
	6 Distribution pipework, etc.	item		
	7 Drainage installation and adaptations.			
	8 Disposal charges (i.e. rates).	week	Time-related charge.	
	9 Disposal costs (i.e. tanker charges).			

1.2.4 Security

Component	Included	Unit	Pricing method	Excluded
1 Security staff	1 Security guards (day and night).	nr (number of staff by number of man hours per week by number of weeks).	Time-related charge.	1 Security staff accommodation (included in section 1.1.2: Site establishment).
	2 Watchmen (day and night).			
2 Security equipment	1 Site pass issue equipment, including maintenance and removal.	item	1 Fixed charge. 2 Time-related charge.	
	2 Site pass consumables.			
	3 CCTV surveillance installation, including maintenance and removal.			
	4 Temporary vehicle control barriers, including maintenance and removal.	nr		
3 Hoardings, fences and gates	1 Perimeter hoardings and fencing and the like to site boundaries and to form site compounds.	m	1 Fixed charge. 2 Time-related charge.	
	2 Access gates, including frames and ironmongery.	nr		
	3 Painting of hoardings, fencing, gates, and the like.	m		
	4 Temporary doors.	nr		
	5 Modification to line of hoarding and fencing during construction.			
	6 Dismantling and removal of hoarding, fencing, gates, and the like.	m		

1.2.5 Safety and environmental protection

Component	Included	Unit	Pricing method	Excluded
1 Safety programme	Works required to satisfy requirements of CDM Regulations: 1 Health and safety manager/officers.	nr (number of staff by number of man hours per week by number of weeks)	Time-related charge.	1 Health and safety manager/officers (included in section 1.2.1: Management and staff). 2 Welfare facilities (included in section 1.2.2: Site establishment).
	2 Safety audits, including safety audits carried out by external consultant.	nr	1 Fixed charge. 2 Time-related charge.	
	3 Staff safety training.	item		
	4 Site safety incentive scheme.			
	5 Notices and information to neighbours.			
	6 Personal protective equipment (PPE), including for employer and consultants.	nr (sets)		
	7 PPE for multi-service gangs.			
	8 Fire points.	nr		
	9 Temporary fire alarms.			
	10 Fire extinguishers.			
	11 Statutory safety signage.	item		
	12 Nurse.	nr (number of staff by number of man hours per week by number of weeks)	Time-related charge.	
	13 Traffic marshals.			
2 Barriers and safety scaffolding	1 Guard rails and edge protection (e.g. to edges of suspended slabs and roofs).	item	1 Fixed charge. 2 Time-related charge.	
	2 Temporary staircase balustrades (i.e. to new staircases during construction).			
	3 Lift shaft protection.			
	4 Protection to holes and openings in ground floor slabs, suspended slabs and the like.			
	5 Debris netting/plastic sheeting.			

Component	Included		Unit	Pricing method		Excluded
	6	Fan protection.	item	1 Fixed charge.		
	7	Scaffold inspections.	nr	2 Time-related charge.		
	8	Hoist run-offs.	item			
	9	Protective walkways.				
	10	Other safety measures.				
3 Environmental protection measures	1	Control of pollution.	item	1 Fixed charge. 2 Time-related charge.		
	2	Residual control of noise.				
	3	Environmental monitoring.				
	4	Environmental manager/consultant.	nr (number of staff by number of man hours per week by number of weeks)	Time-related charge.		
	5	Environmental audits, including safety audits carried out by external consultant.	nr	1 Fixed charge. 2 Time-related charge.		

1.2.6 Control and protection

Component	Included		Unit	Pricing method		Excluded
1 Survey, inspections and monitoring	1	Surveys.	item	1 Fixed charge. 2 Time-related charge.		1 Environmental monitoring (included in section 2.2.5: Safety and environmental protection).
	2	Topographical survey.				
	3	Non-employer dilapidation survey.				
	4	Structural/dilapidations survey adjoining buildings.				
	5	Environmental surveys.				
	6	Movement monitoring.				
	7	Maintenance and inspection costs.				

Component	Included	Unit	Pricing method	Excluded
2 Setting out	1 Setting out primary grids. 2 Grid transfers and level checks. 3 Maintenance of grids. 4 Take over control and independent checks (i.e. on change of subcontractors). 5 Instruments for setting out.	item	1 Fixed charge. 2 Time-related charge.	
3 Protection of works	1 Protection of finished works to project handover. 2 Protection of stairs, balustrades and the like works to project handover. 3 Protection of fittings and furnishings works to project handover. 4 Protection of entrance doors and frames works to project handover. 5 Protection of lift cars and doors works to project handover. 6 Protection of specifically vulnerable products to project handover. 7 Protection of all sundry items.	item	1 Fixed charge. 2 Time-related charge.	
4 Samples	1 Provision of samples. 2 Provision of sample room. 3 Mock-ups and sample panels. 4 Testing of samples/mock-ups, including testing fees. 5 On-site laboratory equipment. 6 Mock-ups of prefabricated units (e.g. residential units, student accommodation units, hotel accommodation and the like).	item	1 Fixed charge. 2 Time-related charge.	
5 Environmental control of building	1 Dry out building. 2 Temporary heating/cooling. 3 Temporary waterproofing, including over roofs. 4 Temporary enclosures.	item	1 Fixed charge. 2 Time-related charge.	

1.2.7 Mechanical plant

Component	Included	Unit	Pricing method	Excluded
1 Generally	Common user mechanical plant and equipment used in construction operations.			Plant and equipment used for specific construction operations, such as: 1 Earthmoving plant. 2 Piling plant. 3 Paving and surfacing plant. 4 Wheel spinners, and road sweepers (included in section 1.2.11: Cleaning). 5 Access scaffolding (included in section 1.2.8: Temporary works).
2 Tower cranes	Type of craneage to be provided shall be stated; with each type separately quantified.			1 Temporary electrical supply to tower crane (included in section 1.2.3: Temporary services).
	1 Hire charges.	week	Time-related charge.	
	2 Crane operator.	week (number of staff by number of man hours per week by number of weeks)		
	3 Overtime for crane and operator.			
	4 Piles for tower crane bases, including maintenance removal.	nr	1 Fixed charge. 2 Time-related charge.	
	5 Temporary bases for tower cranes, including anchors, maintenance; removal and reinstatement on completion (size, in m², to be stated).			
	6 Ties.	week	Time-related charge.	
	7 Connections to temporary electrical supply.	nr	Fixed charge.	
	8 Bring to site, erection, test and commission.			
	9 Periodic safety checks/inspections.	week	Time-related charge.	
	10 Dismantling and removing from site.	nr	Fixed charge.	

Component	Included	Unit	Pricing method	Excluded
	11 Other costs specific to tower crane such as: – chain pack and sundries – relief operator – banksman – man cage.	item	1 Fixed charge. 2 Time-related charge.	
	12 Temporary voids in building structure for craneage, hoists and the like including filling voids after removal.	nr	Fixed charge.	
3 Mobile cranes	Type of craneage to be provided shall be stated; with each type separately quantified.		1 Fixed charge. 2 Time-related charge.	
	1 Mobile crane hire charges, including driver/operator charges.	week		
	2 Attendance.	nr (number of man hours per visit by number of visits)		
	3 Other costs specific to mobile crane hire.	item		
4 Hoists	Type of hoist to be provided shall be stated; with each type separately quantified.			1 Temporary services to hoist installations (included in section 1.2.3: Temporary services).
	1 Goods and passenger hoists, including protection cages and embedment frames.	week	Time-related charge.	
	2 Hoist bases.	nr	1 Fixed charge. 2 Time-related charge.	
	3 Bringing to site, erecting, testing and commissioning.	nr	Fixed charge.	
	4 Dismantling and removing from site.			
	5 Protection systems.	item	1 Fixed charge. 2 Time-related charge.	
	6 Hoist operator, including overtime.	week (number of staff by number of man hours per week by number of weeks)	Time-related charge.	
	7 Beam hoists.	item	1 Fixed charge. 2 Time-related charge.	

Component	Included	Unit	Pricing method	Excluded
	8 Periodic safety checks/inspections.	month	Time-related charge.	
	9 Other costs specific to temporary hoist installations.	item	1 Fixed charge. 2 Time-related charge.	
5 Access plant	1 Fork lifts.	week	1 Fixed charge. 2 Time-related charge.	
	2 Scissor lifts.			
	3 Loading platforms.			
	4 Maintenance of mechanical access equipment.			
	5 Other costs specific to mechanical access equipment.	item		
6 Concrete plant	1 Concrete plant.	week	1 Fixed charge. 2 Time-related charge.	1 Temporary service to concrete plant (included in section 1.2.3: Temporary services).
	2 Plant operator.	week (number of staff by number of man hours per week by number of weeks)	Time-related charge.	
	3 Overtime for plant and operator.			
	4 Bases for concrete plant.	nr	1 Fixed charge. 2 Time-related charge.	
	5 Power connections, including cabling and statutory undertaker's charges for temporary connection to their supply.	nr	1 Fixed charge. 2 Time-related charge.	
	6 Bring to site, erection, test and commission.	nr	Fixed charge.	
	7 Maintenance of concrete plant.	week	Time-related charge.	
	8 Dismantling and removing from site.	nr	Fixed charge.	

Component	Included	Unit	Pricing method	Excluded
7 Other plant	1 Small plant and tools.	week	Time-related charge.	

1.2.8 Temporary works

Component	Included	Unit	Pricing method	Excluded
1 Access scaffolding	Common user access scaffolding (type of access scaffolding to be specified): – access scaffolding to elevations, lift shafts and the like, including: fans and mesh screens. – structural scaffolding (e.g. to party walls). – birdcage scaffolding. – cantilever access scaffolding. – staircase platforms. – primary loading platforms. – travelling access platforms.			1 Scaffolding specific to works packages (included in appropriate element or sub-element). 2 Scaffold inspections (included in sub-element 1.2.5: Safety and environmental protection).
	1 Bringing to site, erecting and initial safety checks.	nr	Fixed charge.	
	2 Hire charges.	week	Time-related charge.	
	3 Altering and adapting during construction.	nr	Fixed charge.	
	4 Dismantling and removing from site.			

Component	Included	Unit	Pricing method	Excluded
2 Temporary works	Common user temporary works: – support scaffolding and propping. – crash decks. – temporary protection to existing trees and/or vegetation. – floodlights.			1 Temporary works design (included in section 1.1.1: Management and staff). 2 Temporary bases, drainage and services to site accommodation (included in section 1.2.2: Site establishment). 3 Temporary roads, paths and pavement, including on-site car parking (included in section 1.2.2: Site establishment (i.e. Builder's work in connection with site accommodation)). 4 Hoardings, fans, fencing and the like to site boundaries and to form site compounds (included in section 1.2.4: Security (hoardings, fences and gates). 5 Temporary earthwork support basement excavations. 6 Temporary props and walings to support contiguous bored pile wall of basement excavations. 7 Traffic management, including traffic marshals and temporary traffic lights (included in section 1.2.5: Safety and environmental protection).
	1 Bringing to site, erecting and initial safety checks.	nr	Fixed charge.	
	2 Hire charges.	week	Time-related charge.	
	3 Altering and adapting during construction.	nr	Fixed charge.	
	4 Dismantling and removing from site.			

1.2.9 Site records

Component	Included	Unit	Pricing method	Excluded
1 Site records	Unless otherwise indicated, costs associated with the following shall be deemed to be included in management and staff costs: 1 Photography: – camera purchase. – consumables. – printing and presentation.	item	1 Fixed charge. 2 Time-related charge.	

Component	Included	Unit	Pricing method	Excluded
	2 Works records: – progress reporting. – site setting out drawings. – condition surveys and reports. – operation and maintenance manuals. – as-built/installed drawings and schedules. – co-ordinating, gathering and compiling health and safety information and presentation to CDM co-ordinator. – compilation of health and safety file (if required).			

1.2.10 Completion and post-completion requirements

Component	Included	Unit	Pricing method	Excluded
1 Testing and commissioning plan	Costs associated with the following shall be deemed to be included in section 1.2.1: Management and staff costs: 1 Preparation of Commissioning Plan.	item	1 Fixed charge. 2 Time-related charge.	1 Testing and commissioning of services.
2 Handover	Unless otherwise indicated, costs associated with the following shall be deemed to be included in section 1.2.1: Management and staff costs: 1 Preparation of Handover Plan. 2 Training of building user's staff in the operation and maintenance of the building engineering services systems. 3 Provision of spare parts for maintenance of building engineering services. 4 Provision of tools and portable indicating instruments for the operation and maintenance of building engineering services systems. 5 Pre-completion inspections. 6 Final inspections.			

Component	Included	Unit	Pricing method	Excluded
3 Post-completion services	1 Supervisory staff (employer/tenant care).	week (number of staff by number of man hours per week by number of weeks)	Time-related charge.	
	2 Handyman.			
	3 Minor materials and sundry items.	item	Fixed charge.	
	4 Insurances.			
	5 Other post-construction staff.	week (number of staff by number of man hours per week by number of weeks)	Time-related charge.	

1.2.11 Cleaning

Component	Included	Unit	Pricing method	Excluded
1 Site tidy	1 Cleaning site accommodation – internal, including cleaning telephone handsets, other office furniture and equipment and window cleaning.	week	Time-related charge.	
	2 Periodic maintenance of site accommodation, including redecoration (internal and external).			
	3 Waste management, including rubbish disposal (including compactor visits; skips and waste bins; roll-off, roll-on waste bins) and other disposal.			
	4 Pest control.			
2 Maintenance of roads, paths and pavings	1 Maintenance of temporary site roads, paths, and pavements.	week	Time-related charge.	
	2 Maintenance of public and private roads, including wheel spinners and road sweepers.			
3 Building clean	1 Final builder's clean.	item	Fixed charge.	

1.2.12 Fees and charges

Component	Included	Unit	Pricing method	Excluded
1 Fees	1 Building control fees, where not paid by the employer. 2 Oversailing fees, where not paid by the employer. 3 Considerate Constructors' Scheme fees (or alternative scheme operated by local authority). 4 Scheme registration fees or similar fees, where not paid by the employer.	item	1 Fixed charge. 2 Time-related charge.	1 Building control fees, where paid by the employer. 2 Oversailing fees, where paid by the employer. 3 Scheme registration fees or similar fees, where paid by the employer.
2 Charges	1 Rates on temporary accommodation.	week	Time-related charge.	1 Statutory undertaker's charges in connection with permanent services to the building. 2 Statutory undertaker's charges in connection with temporary services.
	2 Licences in connection with hoardings, scaffolding, gantries and the like. 3 Licences in connection with crossovers, parking permits, parking bay suspensions and the like.	item	1 Fixed charge. 2 Time-related charge.	

1.2.13 Site services

Component	Included	Unit	Pricing method	Excluded
1 Temporary works	1 Temporary works that are not specific to an element.	item/nr/m/m²/m³	1 Fixed charge. 2 Time-related charge.	1 Temporary screens in connection with minor demolition works and alteration works. 2 Supports to small openings cut into existing walls or after removal of internal walls or the like in connection with minor demolition works and alteration works. 3 Temporary or semi-permanent support for unstable structures or facades – facade retention works (i.e. structures not to be demolished).

Component	Included	Unit	Pricing method	Excluded
2 Multi-service gang	1 Ganger. 2 Labour. 3 Fork lift driver. 4 Service gang plant and transport.	week (number of staff by number of man hours per week by number of weeks)	Time-related charge.	

1.2.14 Insurance, bonds, guarantees and warranties

Component	Included	Unit	Pricing method	Excluded
1 Works insurance	1 Contractor's 'all risks' (CAR) insurance. 2 Contractor's plant and equipment insurance. 3 Temporary buildings insurance. 4 Terrorism insurance. 5 Other insurances in connection with the works. 6 Insurance premium tax (IPT). 7 Allowance for recovery of all or part of insurance premium excess.	item	1 Fixed charge. 2 Time-related charge.	
2 Public liability insurance	1 Non-negligence insurance. 2 Professional indemnity insurance. 3 Insurance premium tax (IPT). 4 Allowance for recovery of all or part of insurance premium excess.	item	1 Fixed charge. 2 Time-related charge.	
3 Employer's (main contractor's) liability insurance	1 Management and staff, including administrative staff. 2 Works operatives. 3 Insurance premium tax (IPT). 4 Allowance for recovery of all or part of insurance premium excess.	item	1 Fixed charge. 2 Time-related charge.	

Component	Included	Unit	Pricing method	Excluded
4 Other insurances	1 Employer's loss of liquidated damages.	item	1 Fixed charge. 2 Time-related charge.	
	2 Latent defects cover.			
	3 Motor vehicles.			
	4 Other insurances.			
	5 Insurance premium tax (IPT).			
	6 Allowance for recovery of all or part of insurance premium excess.			
5 Bonds	1 Tender bonds (if applicable).	item	1 Fixed charge. 2 Time-related charge.	
	2 Performance bonds.			
6 Guarantees	1 Parent company guarantees.	item	1 Fixed charge. 2 Time-related charge.	
	2 Product guarantees, insurance backed guarantees.			
7 Warranties	1 Collateral warranties.	item	1 Fixed charge. 2 Time-related charge.	
	2 Funder's warranties.			
	3 Purchaser's and tenant's warranties.			
	4 Other warranties.			

1 Preliminaries (works package contract)

Part A: Information and requirements

1.1 Project particulars

Sub-heading 1	Sub-heading 2	Information requirements	Supplementary information/notes
1 Project particulars.	1 The project.	Short project title.	
	2 Nature of work package works.	Short description to be stated.	
	3 Location of project.	Full postal address to be stated.	
	4 Length of work package contract/sub-contract.	Period, in weeks, to be stated.	Where to be stated by the contractor, insert 'to be confirmed'.
	5 Names, addresses and points of contact of contractor, employer and consultants.	As for Preliminaries (main contract).	

1.2 Drawings and other documents

Sub-heading 1	Sub-heading 2	Information requirements	Supplementary information/notes
1 Drawings	1 List of drawings from which the bill of quantities was prepared.	As for Preliminaries (main contract).	Exceptions to be stated.
2 Other documents	1 Pre-construction information.	As for Preliminaries (main contract).	

Sub-heading 1	Sub-heading 2	Information requirements	Supplementary information/notes
	2 List of drawings and other documents relating to the work package or sub-contract but not included in the tender documents.	(1) Provide list of drawings and other documents relating to the work package or sub-contract but not included in the tender documents, which may be seen by the contractor during the tender period. (2) Document title, reference, revision, date of issue and author to be stated. (3) Details of where documents can be seen to be stated.	

1.3   The site and existing buildings

Sub-heading 1	Sub-heading 2	Information requirements	Supplementary information/notes
1 The site		(1) As for Preliminaries (main contract).	Cross reference to main contractor's preliminaries.
2 Existing buildings on or adjacent to the site		(2) Additional details relevant to the work package.	
3 Existing mains services	1 On the site.	As for Preliminaries (main contract).	
	2 Adjacent to the site.	As for Preliminaries (main contract).	
4 Health and safety hazards			
5 Site visits		Arrangements for site visits.	Include where information not given in the 'Conditions of Tender'.

1.4   Description of the Works

Sub-heading 1	Sub-heading 2	Information requirements	Supplementary information/notes
1 The Works		Description of the Works (or entire building project).	
2 Preparatory work by others		Description of any work that will be carried out by others under a separate contract before the start of work on site for this contact.	

Sub-heading 1	Sub-heading 2	Information requirements	Supplementary information/notes
	3 Subcontract work 4 Completion work by others	Description for the works comprising the works package.	

1.5 The contract conditions

Sub-heading 1	Sub-heading 2	Information requirements	Supplementary information/notes
1 Conditions of contract	1 [Form of contract title to be stated]	(1) Full title of the standard or bespoke form of contract/sub-contract, including edition, revision, and standard amendments applicable. (2) Reference to any amendments to clauses/conditions to standard form of contract/sub-contract (see note (1)). (3) Reference to any supplementary or special clauses/conditions to standard form of contract/sub-contract.	(1) Where bespoke or uncommon forms of contract are used, a copy is to be appended to the bill of quantities or included as part of the tender.

1.6 Provision, content and use of documents

Sub-heading 1	Sub-heading 2	Information requirements	Supplementary information/notes
1 Definitions and interpretations	1 Definitions. 2 Communication. 3 Products. 4 Site equipment. 5 Drawings. 6 Contractor's choice. 7 Contractor's designed works 8 Submit proposals. 9 Terms used in specification.	As for Preliminaries (main contract).	

Sub-heading 1	Sub-heading 2	Information requirements	Supplementary information/notes
	10 Manufacturer and product references.		
	11 Substitution of products.		
	12 Cross-references.		
	13 Referenced documents.		
	14 Equivalent products.		
	15 Substitution of standards.		
	16 Currency of documents.		
	17 Product sizes.		
2 Documents provided on behalf of employer	1 Additional copies of drawings and documents.	As for Preliminaries (main contract).	
	2 Dimensions.		
	3 Measured quantities		
	4 The specification.		
	5 Divergence from the statutory requirements.		
	6 Employer's policy documents.		
3 Documents provided by the contractor, subcontractors and suppliers	1 Design and production information.	As for Preliminaries (main contract).	
	2 Drawn and other information.		
	3 As-built/installed drawings and information.		
	4 Technical literature»		
	5 Maintenance instructions and guarantees.		
	6 Code for Sustainable Homes.		
	7 Environmental assessment method.		

Sub-heading 1	Sub-heading 2	Information requirements	Supplementary information/notes
Document and data interchange	4 Electronic data interchange (EDI).	As for Preliminaries (main contract).	

1.7 Management of the works

Sub-heading 1	Sub-heading 2	Information requirements	Supplementary information/notes
1 Employer's requirements – generally	1 Insurance.	As for Preliminaries (main contract).	
	2 Professional indemnity insurance	Specific requirements.	
	3 Insurance claims.	As for Preliminaries (main contract).	
2 Programme/progress	1 Programme.	As for Preliminaries (main contract).	
	2 Revised programme.		
	3 Commencement of work on site.		
	4 Work package contractor's progress report.		
3 Cost control	1 Removal/replacement of existing work.	As for Preliminaries (main contract).	
	2 Proposed instructions.		
	3 Measurement of covered work.		
	4 Daywork vouchers.		
	5 Interim valuations and payments.		
	6 Payment for products not incorporated into the works.		
	7 Payment for products stored off-site.		
4 Attendances	1 General attendances	Details of general attendances provided by the main contractor.	
	2 Special attendances.	method of dealing with work package contractor's additional requirements.	

1.8 Quality standards and quality control

Sub-heading 1	Sub-heading 2	Information requirements	Supplementary information/notes
1 Standards of products and executions	1 Incomplete information.	As for Preliminaries (main contract).	
	2 Workmanship skills.		
	3 Quality of products.		
	4 Quality of execution.		
	5 Compliance.		
	6 Inspections		
	7 Manufacturer's recommendations/instructions.		
2 Samples/approvals	1 Samples.	As for Preliminaries (main contract).	
3 Accuracy/setting out	1 Accuracy of instruments.	As for Preliminaries (main contract).	
	2 Setting out.		
	3 Appearance and fit.		
	4 Critical dimensions.		
	5 Levels of structural floors.		
	6 Record drawings.		
4 Services	1 Services regulations.	As for Preliminaries (main contract).	
	2 Water regulations/byelaws notification.		
	3 Water regulations/byelaws contractor's certificate.		
	4 Electrical installation certificate.		
	5 Gas, oil and solid fuel appliance installation certificate.		
	6 Mechanical and electrical services.		
5 Supervision/inspection/ defective work	1 Supervision.	As for Preliminaries (main contract).	
	2 Defects in existing work.		

Sub-heading 1	Sub-heading 2	Information requirements	Supplementary information/notes
	3 Proposals for rectification of defective products/executions.		
	4 Proposals for rectification of defective products/executions.		
	5 Quality control.		
6 Work at or after completion	1 Commissioning and testing.	Specific requirements.	

1.9 Security, safety and protection

Sub-heading 1	Sub-heading 2	Information requirements	Supplementary information/notes
1 Security/health and safety	1 Pre-construction information.	As for Preliminaries (main contract).	
	2 Execution hazards.		
	3 Product hazards.		
	4 Occupied premises.		
	5 Passes.		
	6 Occupier's rules and regulations.		
	7 Use of mobile telephones.		
	8 Working precautions/ restrictions.		
2 Protection against	1 Noise control.	As for Preliminaries (main contract).	
	2 Pollution control.		
	3 Fuels, lubricants and hydraulic fluids.		
	4 Nuisance.		
	5 Asbestos containing materials (ACM)s.		
	6 Antiquities.		
	7 Fire prevention.		
	8 Smoking on site.		
	9 Moisture.		

Sub-heading 1	Sub-heading 2	Information requirements	Supplementary information/notes
	10 Infected timber/contaminated materials.	As for Preliminaries (main contract).	
	11 Waste.		
	12 Electromagnetic interference.		
3 Protection	1 Existing features.	As for Preliminaries (main contract).	
	2 Existing work.		
	3 Building interiors.		
	4 Existing structures.		
	5 Materials for recycling and/or reuse.		
	6 Protection of work package contractor's work.	Specific requirements.	

1.10 Specific limitations on method, sequence and timing

Sub-heading 1	Sub-heading 2	Information requirements	Supplementary information/notes
1 Generally		As for Preliminaries (main contract).	
2 Use or disposal of materials			
3 Working hours		Specific requirements.	

1.11 Site accommodation/services/facilities/temporary works

Sub-heading 1	Sub-heading 2	Information requirements	Supplementary information/notes
1 Services and facilities	1 Lighting and power.	As for Preliminaries (main contract).	
	2 Gas.		
	3 Water.		
	4 Mobile telephones.		

Sub-heading 1	Sub-heading 2	Information requirements	Supplementary information/notes
	5 Temperature and humidity.		
	6 Beneficial use of permanent installed systems.		
	7 Meter readings.	Specific requirements.	
	8 Other requirements.	Specific requirements.	
2 Temporary works	1 Name boards.		
	2 Advertising.		
	3 Other requirements.		

1.12 Operation/maintenance of finished building

Sub-heading 1	Sub-heading 2	Information requirements	Supplementary information/notes
1 Operation and maintenance manual	1 Generally.	As for Preliminaries (main contract).	The operation and maintenance information, the health and safety file, and all other information can be combined as a single document. In this case, the document can be referred to as the 'building manual'.
	2 Content.	Specify information and documents to be provided.	
2 Operation and maintenance manual information			
3 Health and safety file	1 Generally.	As for Preliminaries (main contract).	
	2 Content.	Specify information and documents to be provided.	
4 Health and safety information		Specify information and documents to be provided.	
5 Other information		Specific requirements.	

1 Preliminaries (works package contract)

Part B: Pricing schedule

1.1 Management and staff

Component	Included	Unit	Pricing method	Excluded
1 Project specific management and staff	Work package contractor's project-specific management and staff.	week (number of staff by number of man hours per week by number of weeks)		
2 Staff travel	Costs associated with off-site visits such as:			
	1 Visits to employer's and consultant's offices.	nr (number of occasions)	Fixed charge.	
	2 Visits to main contractors' offices/works.			
	3 Overseas visits.			
	4 Accommodation charges and overnight expenses.			

1.2 Site establishment

Component	Included	Unit	Pricing method	Excluded
1 Site accommodation	Work package contractor's project-specific site accommodation.			
	Type and extent of accommodation to be provided to be stated; with each type separately quantified..			
	1 Purchase charges.	item	Fixed charge.	
	2 Hire charges.	week	Time-related charge.	
	3 Delivery of temporary site accommodation to site, erection, construction and removal.	item	Fixed charge.	
	4 Intruder alarms.			
	5 Land/property rental where site accommodation located off-site.	week	Time-related charge.	

TABULATED WORK SECTIONS

Component	Included	Unit	Pricing method	Excluded
	6 Alterations and adaptations to site accommodation, including partitioning, doors, painting and decorating, and the like.	item	Fixed charge.	
	7 Relocation and alterations of temporary accommodation during construction stage.			
	8 Removal of site accommodation and temporary works in connection with site accommodation.			
2 Temporary works in connection with site establishment	1 Temporary bases and foundations for site accommodation, including maintenance and reinstatement of existing surfaces on completion of the works.	m²	1 Fixed charge. 2 Time-related charge.	
	2 Connections to temporary service, including maintenance and removal on completion of the works	nr		
	3 Connections to temporary drainage, in including maintenance and removal on completion of the works.			
3 Furniture and equipment	1 Workstations for staff.	nr	1 Fixed charge. 2 Time-related charge.	
	2 General office furniture.	item		
	3 Floor coverings.			
	4 Heaters, including maintenance of heaters.			
	5 Other office equipment, including maintenance.			
	6 Removal of furniture and equipment.			
	7 Maintenance of furniture and floor covering.			

Component	Included	Unit	Pricing method	Excluded
4 IT systems	1 Computer hardware, including purchase/rental, installation, initial set up, maintenance and running costs, such as: – desktop computers and laptop computers – CAD stations – server and network equipment – printers and plotters – other computer system/hardware.	item	1 Fixed charge. 2 Time-related charge.	
	2 Software and software licences.			
	3 Modem lines, modems and connections (i.e. email and internet capability).			
	4 WAN lines and connections (if on WAN).			
	5 Line rental charges.	week	Time-related charge.	
	6 Internet/website addresses.	nr	Fixed charge.	
	7 Internet service provider (ISP) charges.			
	8 Line calls charges.	week	Time-related charge.	
	9 IT support and maintenance.			
5 Consumables and services	1 Stationery.	week	Time-related charge.	
	2 Computer and printer consumables (e.g. ink cartridges).			
	3 Postage.			
	4 courier charges.			
	5 Tea, coffee, water bottles, etc.			
	6 First aid consumables.			
	7 Photocopier consumables (e.g. paper and toners).			
	8 Fax consumables (e.g. paper and toners).			
	9 Drawing printer consumables (e.g. ink cartridges).			

Component	Included	Unit	Pricing method	Excluded
6 Sundries	1 Work package contractor's signboards.	item	Fixed charge.	

1.3 Temporary services

Component	Included	Unit	Pricing method	Excluded
1 Temporary water supply	1 Temporary connections. 2 Distribution equipment, installation and adaptations. 3 Meter charges.	nr item week	1 Fixed charge. 2 Time-related charge.	
2 Temporary gas supply	1 Gas connection. 2 Distribution equipment, installation and adaptations. 3 Charges. 4 Bottled gas.	nr item week	1 Fixed charge. 2 Time-related charge.	
3 Temporary electricity supply	1 Temporary connections.	nr	Fixed charge.	
4 Temporary telecommunication system	1 Landlines (including connection and rental charges), including: – telephone and fax lines – ISDN lines. 2 Telephone and facsimile equipment (including connection and rental charges), including: – PABX equipment – handsets, including purchase or rental – fax machines, including purchase or rental – installation of equipment – maintenance of equipment. 3 Mobile (cellular) phones, including: – mobile phones, including purchase or rental and connection charges – spare batteries – mobile phone charges .	item	1 Fixed charge. 2 Time-related charge.	1 Fax consumables (included in section 1.2.5: Consumables and services).

Component	Included	Unit	Pricing method	Excluded
	4 Telephone charges, including: – telephone call charges – fax charges – fax and telephone consumables. 5 Radios (including purchase or rental charges), including: – base set – handsets and chargers – repairs and maintenance – licences – spare batteries.			

1.4 Control and protection

Component	Included	Unit	Pricing method	Excluded
1 Survey, inspections and monitoring	1 Surveys. 2 Topographical survey. 3 Non-employer dilapidation. survey 4 Structural/dilapidations survey adjoining buildings. 5 Environmental surveys. 6 Movement monitoring. 7 Maintenance and inspection costs.	item	1 Fixed charge 2 Time-related charge.	1 Environmental monitoring (included in section 1.2.5: Safety and environmental protection)
2 Setting out	1 Setting out primary grids. 2 Grid transfers and level checks. 3 Maintenance of grids. 4 Take over control and independent checks (i.e. on change of subcontractors). 5 Instruments for setting out.	item	1 Fixed charge. 2 Time-related charge.	

Component	Included	Unit	Pricing method	Excluded
3 Protection of Works	1 Protection of finished works to project handover.	item	1 Fixed charge. 2 Time-related charge.	
	2 Protection of stairs, balustrades and the like to project handover.			
	3 Protection of fittings and furnishings to project handover.			
	4 Protection of entrance doors and frames to project handover.			
	5 Protection of lift cars and doors to project handover.			
	6 Protection of specifically vulnerable products to project handover.			
	7 Protection of all sundry items.			
4 Samples	1 Provision of samples.	item	1 Fixed charge. 2 Time-related charge.	
	2 Provision of sample room.			
	3 Mock-ups and sample panels.			
	4 Testing of samples/mock-ups, including testing fees.			
	5 On-site laboratory equipment.			
	6 Mock-ups of complete units (e.g. residential units, student accommodation units, hotel accommodation and the like).			
5 Environmental control of building	1 Dry out building.	item	1 Fixed charge. 2 Time-related charge.	
	2 Temporary heating/cooling.			
	3 Temporary waterproofing, including over roofs.			
	4 Temporary enclosures.			

1.5 Mechanical plant

Component	Included	Unit	Pricing method	Excluded
1 Mechanical plant	Type of plant to be provided shall be stated; with each type separately quantified.			
	1 Bases.	nr	1 Fixed charge. 2 Time-related charge.	
	2 Bringing to site, erecting, testing and commissioning.	nr	Fixed charge.	
	3 Dismantling and removing from site.			
	4 Protection systems.	item	1 Fixed charge. 2 Time-related charge.	
	5 Operator/driver; including overtime.	week (number of staff by number of man hours per week by number of weeks)	Time-related charge.	
	6 Periodic safety checks/inspections.	month	Time-related charge.	
	7 Other costs' specific charges.	item	1 Fixed charge. 2 Time-related charge.	

1.6 Temporary works

Component	Included	Unit	Pricing method	Excluded
1 Access scaffolding	Access scaffolding specifically required by work package contractor (type of access scaffolding to be specified): – access scaffolding to elevations, lift shafts and the like, including: fans and mesh screens. – structural scaffolding (e.g. to party walls). – birdcage scaffolding. – cantilever access scaffolding. – staircase platforms. – primary loading platforms. – travelling access platforms.			
	1 Bringing to site, erecting and initial safety checks.	nr	Fixed charge.	
	2 Hire charges.	week	Time-related charge.	
	3 Altering and adapting during construction.	nr	Fixed charge.	
	4 Dismantling and removing from site.			

Component	Included	Unit	Pricing method	Excluded
2 Temporary works	Temporary works specifically required by work package contractor: – support scaffolding and propping – crash decks – temporary protection to existing trees and/or vegetation – floodlights.			
	1 Bringing to site, erecting and initial safety checks.	nr	Fixed charge.	
	2 Hire charges.	week	Time-related charge.	
	3 Altering and adapting during construction.	nr	Fixed charge.	
	4 Dismantling and removing from site.			

1.7 Site records

Component	Included	Unit	Pricing method	Excluded
1 Site records	Unless otherwise indicated, costs associated with the following shall be deemed to be included in management and staff costs: 1 Photography: – Camera purchase – Consumables – Printing and presentation	item	1 Fixed charge. 2 Time-related charge.	

Component	Included	Unit	Pricing method	Excluded
	2 Works records: – Progress reporting – Site setting out drawings – Condition surveys and reports – Operation and maintenance manuals – as-built/installed drawings and schedules – co-ordinating, gathering and compiling health and safety information and presentation to CDM co-ordinator – compilation of health and safety file (if required).			

1.8 Completion and post-completion requirements

Component	Included	Unit	Pricing method	Excluded
1 Testing and commissioning plan	Costs associated with the following shall be deemed to be included in section 1.1: Management and staff costs: 1 Preparation of commissioning plan.	item	1 Fixed charge. 2 Time-related charge.	1 Testing and commissioning of services
1 Handover	Unless otherwise indicated, costs associated with the following shall be deemed to be included in section 1.1: Management and staff costs: 1 Preparation of Handover Plan. 2 Training of building user's staff in the operation and maintenance of the building engineering services systems. 3 Provision of spare parts for maintenance of building engineering services. 4 Provision of tools and portable indicating instruments for the operation and maintenance of building engineering services systems. 5 Pre-completion inspections. 6 Final inspections.	item	1 Fixed charge. 2 Time-related charge.	

Component	Included	Unit	Pricing method	Excluded
3 Post-completion services	1 Supervisory staff (employer/tenant care).	week (number of staff by number of man hours per week by number of weeks)	Time-related charge.	
	2 Handyman.			
	3 Minor materials and sundry items.	item	Fixed charge.	
	4 Insurances.			
	5 Other post-construction staff.	week (number of staff by number of man hours per week by number of weeks)	Time-related charge.	

1.9 Cleaning

Component	Included	Unit	Pricing method	Excluded
1 Site tidy	1 Cleaning site accommodation – internal, including cleaning telephone handsets, other office furniture and equipment and window cleaning.	week	Time-related charge.	
	2 Periodic maintenance of site accommodation, including redecoration (internal and external).			
	3 Waste management, including rubbish disposal (including compactor visits; skips and waste bins; roll-off, roll-on waste bins) and other disposal.			
	4 Pest control.			

1.10   Fees and charges

Component	Included	Unit	Pricing method	Excluded
1 Charges	1 Rates on temporary accommodation.	week	Time-related charge.	

1.11   Insurances, bonds, guarantees and warranties

Component	Included	Unit	Pricing method	Excluded
1 Insurances		item	1 Fixed charge. 2 Time-related charge.	

2   Off-site manufactured materials, components or buildings

Drawings that must accompany this section of measurement.		Mandatory information to be provided.	1 Site plans. 2 Plans. 3 Sections. 4 Elevations. 5 Installation details.	Notes, comments and glossary
Minimum information that must be shown on the drawings that accompany this section of measurement.		Works and materials deemed included.	1 Major dimensions of component, structure or unit. 2 Location of component, structure or unit.	1 Kind and quality of materials. 2 Method of fixing or installing. 3 Connecting to other work and services. 4 Special finishes. 1 All factory applied finishes. 2 Transport from factory to site. 3 Offloading and storing on site. 4 Setting, hoisting and placing in final position. 5 All connection and joint materials. 6 All service connections. 7 Disposal of all packaging and protective materials.	
Item or work to be measured	Unit	Level one	Level two	Level three	Notes, comments and glossary
1 Component	nr	1 Overall dimensions.	1 Description of component.	1 Method of fixing/installing. 2 Height above structural ground floor level. 3 Services connections.	1 These are prefabricated proprietary components that are not adequately covered by the other Work Sections in this document.

Item or work to be measured	Unit	Level one	Level two	Level three	Notes, comments and glossary
2 Prefabricated structures	nr	1 Overall dimensions.	1 Roofs. 2 External walls. 3 Internal walls/partitions. 4 Floors. 5 Stairs. 6 Bridges. 7 Masts. 8 Other; type stated.		1 These are complete or substantially complete building elements of proprietary construction, largely prefabricated. The fixing of items supplied only as part of the proprietary package is included here. Other work not forming part of the proprietary package is measured separately in the appropriate Work Section.
3 Prefabricated building units	nr		1 Toilet/bathroom units. 2 Sound proof rooms. 3 Cold rooms. 4 Spray booths. 5 Kiosks. 6 Other; type stated.		1 These are complete or substantially complete room units, usually of proprietary construction, for incorporation into buildings, structures or siteworks. The list is not exhaustive.
4 Prefabricated buildings	nr		1 Description of building.		1 These are complete or substantially complete building superstructures of proprietary construction, largely prefabricated. The fixing of items supplied only as part of the proprietary package is included here. Other work not forming part of the proprietary package is measured separately in the appropriate Work Section.

3 Demolitions

Demolitions
Shoring, facade retention and temporary works

		Mandatory information to be provided.	Notes, comments and glossary
Drawings that must accompany this section of measurement.	1 Location drawings. 2 Drawings of existing structures to show full extent of demolition.	1 Brief description and size of structure to be demolished. 2 Any limitations due to presence of toxic or hazardous materials. 3 Extent of parts of structure to be temporarily retained. 4 Asbestos surveys.	
Minimum information that must be shown on the drawings that accompany this section of measurement.	1 Lowest level of demolition. 2 Extent of any temporary works not at the discretion of the contractor. 3 Position and extent of any temporary screens, roofs and the like.	Works and materials deemed included. 1 All temporary works unless stated otherwise. 2 Temporarily diverting, maintaining or sealing off existing services. 3 Disposal of all debris unless stated otherwise. 4 Method of demolition unless stated otherwise. 5 All temporary support left to the discretion of the contractor. 6 Clearing away all temporary works. 7 Disposing of rainwater. 8 Making good all work disturbed.	1 The lowest level will include basements. 2 If a floor slab is to be removed then the lowest level must be stated as to the underside of that slab.

Item or work to be measured	Unit	Level one	Level two	Level three	Notes, comments and glossary
1 Demolitions	item	1 All structures. 2 Individual structures. 3 Parts of structures.	1 Description of building(s) or parts of building(s). 2 Lowest level to which structure(s) to be demolished.	1 Limitations on disposal of materials. 2 Any material to be retained for re-use. 3 Any material to remain the property of the employer.	1 Contractor should be advised to inspect the structure(s) to be demolished.
2 Temporary support of structures, roads and the like	item	1 Parts of existing building to be retained. 2 Adjoining buildings not forming part of the works. 3 Roads and other surfaces to be retained. 4 Any other existing feature to be retained.	1 Description of building(s) or parts of building(s) or road or other surface or feature to be retained. 2 Type of shoring. 3 Length of exposed edge of surface to be retained and average height(s).		1 Support is for parts of the structure that must be retained. It does not mean any type of support required as incidental to the demolitions.
3 Temporary works	m²	1 Roofs. 2 Screens. 3 Floors. 4 Roads.	1 Weatherproof. 2 Watertight. 3 Dustproof. 4 Fireproof. 5 Any other requirement: type stated.	1 Method of construction if not at the discretion of the contractor. 2 Maintaining: duration stated. 3 Adapting during course of works. 4 Clearing away. 5 Disposing of rainwater: details stated. 6 Providing openings: details stated.	1 In order to ensure the full extent and scope of this work the surveyor may need to provide additional information if not readily ascertained from the drawings.

Item or work to be measured	Unit	Level one	Level two	Level three	Notes, comments and glossary
4 Decontamination	item	1 Removal of hazardous materials. 2 Decontamination of existing premises. 3 Infestation removal.	1 Scope of work. 2 Type of contamination. 3 Prior to demolition. 4 During demolition or repair process.		
5 Recycling	item	1 Detailed description of type of material to be recycled and any limitations imposed by employer or local authority.	1 To be collected by local authority 2 To be transported to recycling depot, details and location stated.		

4 Alterations, repairs and conservation

Alteration work to existing buildings
Repairs/cleaning/renovating and conserving
Decontamination
Re-cycling

Drawings that must accompany this section of measurement.		Mandatory information to be provided.	Notes, comments and glossary
1	Location drawings.	1 Description of operations where not left to discretion of the contractor.	
2	Drawings of existing structures.	2 Specific location of each item of work relative to the existing building.	
		3 Details of all materials to be set aside for subsequent re-use including means of storage.	
		4 Details of all materials to remain the property of the employer including means of storage.	
		5 Any restrictions on method, sequence and/or timing of the works.	
		6 Any restrictions on methods of storage of materials to be re-used or to remain property of the employer.	
		7 Any restrictions on method or location of disposal of waste.	
		8 Compliance with all regulations relating to the handling, transport and disposal of hazardous waste materials.	
		9 Asbestos surveys.	

Minimum information that must be shown on the drawings that accompany this section of measurement.

1 Scope and location of work relative to existing structures.

Works and materials deemed included.

1 All temporary works including shoring and scaffolding incidental to the work excluding those listed below.
2 Making good all work disturbed.
3 Extending and making good existing finishes.
4 Disposal of all waste materials.
5 All work and materials incidental to the items of alteration.
6 Materials required for bonding new work to existing.

Notes, comments and glossary

1 The rules within this section apply to works to existing buildings as defined in clause 3.3.4.1 of the Measurement Rules for Building Works.
2 Inserting new work includes re-fixing or re-using removed materials.
3 All materials arising from these works become the property of the contractor unless otherwise stated.
4 The rules within this section do not apply to temporary works except as those listed in Rule 24 of this section.

Item or work to be measured	Unit	Level one	Level two	Level three	Notes, comments and glossary
1 Works of alteration	item	1 Dimensioned description sufficient to identify extent and location of work.	1 Extent, nature and scope of work described including type and thickness of existing structure.		1 Details must be given of all work involved in each item including method of operation where not at the discretion of the contractor.
2 Removing	item m² m nr	1 Fittings and fixtures. 2 Plumbing items or installations. 3 Electrical items or installations. 4 Finishes. 5 Coverings. 6 Pavings.	1 Details sufficient for identification to be stated. 2 Approximate area or size of area of each type of finish, covering or paving.		1 Disconnecting and, if required, subsequent re-connection of plumbing and electrical or other services installations is deemed included. 2 The surveyor shall choose the unit most suitable for the type of work being removed.
3 Cutting or forming openings 4 Cutting or forming recesses 5 Cutting back 6 Filling in openings 7 Filling in recesses	item m² m nr	1 Dimensioned description. 2 Type and thickness of existing structure. 3 Method of performing the work if not left to discretion of the contractor.	1 Re-use of existing materials stated. 2 Type and size(s) of new materials stated.		1 Details given of new work are to be the equivalent of those details required by the rules for the measurement of the same in other Work Sections. 2 The surveyor shall choose the unit most suitable for the type of work being cut or filled.

Item or work to be measured	Unit	Level one	Level two	Level three	Notes, comments and glossary
8 Removing existing and replacing	item m² m nr	1 Thickness stated. 2 Width and thickness stated. 3 Length, width and thickness stated. 4 Dimensioned description.	1 Brickwork. 2 Concrete. 3 Stonework. 4 Timber. 5 Glass. 6 Other, type stated.	1 Treatment of exposed sound surface(s) stated. 2 Treatment of exposed reinforcement or other material stated. 3 Making good with new materials other than to match existing to be described. 4 Bonding new to existing.	1 Formwork and any other form of temporary support is deemed included. 2 The unit of measurement shall be left to the discretion of the surveyor but shall reflect the size and extent of the work. 3 Making good is deemed to be to match existing unless described otherwise.
9 Preparing existing structures for connection or attachment of new work	nr	1 Dimensioned description.	1 Description sufficient to determine scope and location.	1 Nature of existing structure to receive new work.	1 This will include preparing structural steel sections for attachment to new steel framing and the like.
10 Repairing	m² m nr item	1 Thickness stated. 2 Width and thickness stated. 3 Length, width and thickness stated. 4 Dimensioned description.	1 Nature of surface to be repaired stated.	1 Method of repairing stated.	1 The unit of measurement shall be left to the discretion of the surveyor but shall reflect the size and extent of the work.
11 Repointing joints	m	1 Nature of existing joint. 2 Type of pointing.	1 Materials required stated.	1 Width and depth stated.	1 Removal of existing joint material and preparation of exposed surfaces deemed included.

Item or work to be measured	Unit	Level one	Level two	Level three	Notes, comments and glossary
12 Repointing	m² m nr	1 Thickness stated. 2 Width and thickness stated. 3 Length, width and thickness stated.	1 Nature of existing surface. 2 Size of existing components. 3 Bond of existing joints. 4 Type of new pointing. 5 Width and depth of raking out of existing joints.	1 Composition and mix of mortar and/or other joint material(s).	1 Types of surface would include brickwork, blockwork, stonework and the like. 2 Areas each less than 1m² to be enumerated. 3 Linear items would include reveals, wall ends and the like. 4 The unit of measurement shall be left to the discretion of the surveyor but shall reflect the size and extent of the work.
13 Resin or cement impregnation/injection	m² m nr item	1 Thickness stated. 2 Width and thickness stated. 3 Length, width and thickness stated. 4 Dimensioned description.	1 Method of impregnation or injection stated. 2 Nature of existing material. 3 Nature of existing finish where applicable. 4 Thickness or depth of treatment.	1 Centres or spacings of drilling holes. 2 Localised removal of finishes.	1 Work is deemed to include making good holes and finishes on completion. 2 Overall removal of finishes prior to this work would be measured elsewhere. 3 The unit of measurement shall be left to the discretion of the surveyor but shall reflect the size and extent of the work.
14 Inserting new walls ties	m² nr	1 Size and type of new ties.	1 Method of insertion. 2 Nature and thickness of outer skin. 3 Nature of existing finish where applicable.	1 Centres or spacings of drilling holes. 2 Localised removal of finishes.	1 Work is deemed to include making good holes and finishes on completion. 2 Overall removal of finishes prior to this work would be measured elsewhere.
15 Re-dressing existing flashings and the like	m nr	1 Girth and thickness stated. 2 Length, width and thickness stated.	1 Dimensioned description of flashing. 2 Description of new profile.	1 Raking out existing joint. 2 Repointing with new material: method and type of pointing material stated.	1 The girths and lengths stated are net. No allowance to be made for additional materials required for labours. 2 Removal, cleaning, re-shaping, trimming and re-fixing existing flashing is deemed included.

Item or work to be measured	Unit	Level one	Level two	Level three	Notes, comments and glossary
16 Damp-proof course renewal 17 Damp-proof course insertion	m	1 Method of renewal or insertion stated. 2 Nature and thickness of existing wall. 3 Nature of existing finishes where applicable. 4 Thickness or depth of treatment.	1 Centres or spacings of drilling holes. 2 Localised removal of finishes.	1 Chemical. 2 Injection mortar. 3 Electro osmosis. 4 Other mechanical methods.	1 Work is deemed to include making good holes and finishes on completion. 2 Overall removal of finishes prior to this work would be measured elsewhere.
18 Cleaning surfaces 19 Removing stains 20 Artificial weathering	m² ---- m ---- nr ---- item	1 Over 500mm wide. ---- 2 Not exceeding 500mm wide: width stated. ---- 3 Length and width stated. ---- 4 Dimensioned description.	1 Nature of surface to be treated stated. 2 Required finished appearance.	1 Treatment material stated. 2 Method of treating stated.	1 The unit of measurement shall be left to the discretion of the surveyor. 2 Any repair and remedial works required to the surface prior to treatment to be measured separately.
21 Renovating	m² ---- m ---- nr ---- item	1 Thickness stated. ---- 2 Width and thickness stated. ---- 3 Length, width and thickness stated. ---- 4 Dimensioned description.	1 Brickwork. 2 Concrete. 3 Stonework. 4 Timber. 5 Other: type stated.	1 Details and nature of the renovation stated. 2 Materials required stated. 3 Method of renovation stated where not at the discretion of the contractor.	1 The unit of measurement shall be left to the discretion of the surveyor but shall reflect the size and extent of the work.
22 Conserving	m² ---- m ---- nr ---- item	1 Thickness stated. ---- 2 Width and thickness stated. ---- 3 Length, width and thickness stated. ---- 4 Dimensioned description.	1 Brickwork. 2 Concrete. 3 Stonework. 4 Timber. 5 Other: type stated.	1 Details and nature of the conservation stated. 2 Materials required stated. 3 Method of conservation stated where not at the discretion of the contractor.	1 The unit of measurement shall be left to the discretion of the surveyor but shall reflect the size and extent of the work.

Item or work to be measured	Unit	Level one	Level two	Level three	Notes, comments and glossary
23 Decontamination	item	1 Removal of toxic/hazardous materials. 2 Decontamination of existing premises. 3 Infestation removal/eradication. 4 Fungus removal/eradication.	1 Scope and location of work. 2 Type of contamination/infestation/fungus to be treated.	1 Preparatory works.	1 Sufficient information must be given to fully describe the cause of the contamination or infestation together with such information as is appropriate to allow the contractor to fully treat the condition described. 2 (Excluding decontamination of existing ground: this work is measured in Work Section 5: Excavating and Filling.)
24 Temporary works Roads.	m²	1 Roofs. 2 Screens. 3 Floors.	1 Weatherproof. 2 Watertight. 3 Dustproof. 4 Fireproof. 5 Any other requirement: type stated.	1 Method of construction if not at the discretion of the contractor. 2 Maintaining: duration stated. 3 Adapting during course of works. 4 Clearing away. 5 Disposing of rainwater: details stated. 6 Providing openings: details stated.	1 In order to ensure the full extent and scope of this work the surveyor may need to provide additional information if not readily ascertained from the drawings.
25 Recycling	item	1 Detailed description of type of material to be recycled and any limitations imposed by employer or local authority.	1 To be collected by local authority. 2 To be transported to recycling depot, details and location stated.		

5 Excavating and filling

Site clearance/preparation
Excavations
Disposal
Fillings
Membranes

Drawings that must accompany this section of measurement.	Mandatory information to be provided.	Notes, comments and glossary
1 Site plan(s) showing all major excavations. 2 Locations of spoil heaps if not left to the discretion of the contractor. 3 Existing site survey.	1 Location of works in accordance with clause 3.3.3.1 of the Measurement Rules for Building Works. 2 Date of existing site survey. 3 Ground conditions including anticipated stability of excavations. 4 Ground water level(s) and date(s) established. 5 Nature of any known hazardous contamination on the site or in the ground including any restrictions on disposal of surface or ground water. 6 Starting level of each type of excavation. 7 Levels of rock where applicable.	1 Quantities relate only to the depths and dimensions shown on the drawings and overdig is not included. This applies to associated Extra over items and backfilling.

Minimum information that must be shown on the drawings or any other documents that accompany this section of measurement.	Works and materials deemed included.
1 Original and proposed ground levels. 2 Any item(s) that must remain on site during the works and be protected from damage. 3 Any item(s) adjacent to site that may impact on the works. 4 Details of trial pits or boreholes including their location. 5 Details of live over or underground services including their location. 6 Pile sizes and their locations where applicable.	1 Disposal of all surface water. 2 Support to faces of excavation unless not at the discretion of the contractor. 3 Working space. 4 Excavation and filling for temporary works unless not at the discretion of the contractor. 5 Levelling, grading, trimming and compacting surfaces exposed by the excavations. 6 Curved work. 7 Multiple handling of excavated materials on site unless specified. 8 All excavated material is deemed to be inert unless described otherwise.

| | | | | | 1 The ground water level must be re-established at the time each excavation is carried out and is defined as the post-contract ground water level.
2 Ground water levels subject to periodic changes due to tidal or similar effects are to be so described giving the mean high and low water levels.
3 The quantities given for excavation and filling are the bulk before excavating or the net void to be filled. No allowance is made for subsequent variations to bulk or for the extra space taken up by working space or earthwork support unless the type of backfill is not left to the discretion of the contractor. This also applies to the quantities given for any subsequent Extra over items listed in Rule 7. |

Item or work to be measured	Unit	Level one	Level two	Level three	Notes, comments and glossary
1 Preliminary sitework	item nr	1 Locating underground services. 2 Trial pits to locate existing services or determine ground conditions.	1 Maximum depth and type of service stated.	1 Specified location(s) stated. 2 Means of locating service if not left to discretion of contractor.	
	nr	3 Boreholes to determine ground conditions.	1 Diameter. 2 Maximum depth stated.	1 Destination of core samples stated. 2 Type and extent of report required.	

Item or work to be measured	Unit	Level one	Level two	Level three	Notes, comments and glossary
2 Removing trees 3 Removing tree stumps	nr	1 Girth 500mm to 1,500mm. 2 Girth 1,500 to 3,000mm. 3 Girth over 3,000mm, stated in 1,500mm stages.	1 Filling material stated.		1 Removing trees is deemed to include removing the stump and roots unless otherwise stated. 2 Tree girths are measured at a height of 1.00m above original ground level. 3 Stump girths are measured at the top. 4 This work is deemed to include: grubbing up roots, disposal off site of all material arising, and filling voids.
4 Site clearance	m²	1 Clear site of all vegetation and other growth and dispose off site.	1 Description sufficient to identify scope and location of work.		1 All growth includes trees and tree stumps less than 500mm girth, bushes, scrub, hedges and the like unless specifically designated to remain. 2 The removal of invasive vegetation such as Japanese knotweed or the like should be specifically mentioned in the description.
5 Site preparation	m²	1 Lifting turf for preservation: thickness stated.	1 Method and location of preservation stated.		
	m²	2 Remove topsoil: depth stated.			
	m²	3 Remove hard surface paving; thickness stated.	1 Destination stated. 2 Type of paving stated.	1 Method of breaking up if not left to discretion of contractor.	1 Excludes any hardcore beds below the pavings. 2 Removal of the hardcore is treated as reduced level excavation.

Item or work to be measured	Unit	Level one	Level two	Level three	Notes, comments and glossary
5 Site preparation—cont.	nr	4 Remove specific items.	1 Dimensioned description sufficient to identify size and location of each item.		1 Any existing items on site not specifically designated to remain including all types of rubbish such as abandoned cars, fridges and the like. 2 This excludes all but the simplest of building structures whose demolition is covered in Work Section 3: Demolitions. 3 Removal of any associated foundations, fixings, supports, fastenings and the like is deemed included.
6 Excavation, commencing level stated if not original ground level	m³	1 Bulk excavation.	1 Not exceeding 2m deep. 2 Over 2m not exceeding 4m deep. 3 And thereafter in stages of 2m.	1 Details of obstructions in ground to be stated.	1 Bulk excavation includes excavating to reduce levels or to form basements, pools, ponds or the like. For clarity each type of excavation may be measured and described separately. 2 Obstructions will be piles, manholes and the like that must remain undisturbed.
	m³	2 Foundation excavation.			1 Foundation excavation includes excavating for strip and pad foundations, pile caps and all other types of foundations. 2 For clarity each type of excavation may be measured and described separately.

5 EXCAVATING AND FILLING

Item or work to be measured	Unit	Level one	Level two	Level three	Notes, comments and glossary
7 Extra over all types of excavation irrespective of depth	m³	1 Excavating in.	1 Hazardous material, details stated. 2 Non-hazardous material, details stated. 3 Below ground water level. 4 In running water. 5 Unstable ground.		1 Hazardous materials are any material that require special precautions taken when handling, transporting or disposing. 2 Ground water is any water encountered below the established water table level. It does not include water from underground streams, broken drains or culverts or water arising from surface flooding. 3 Running water is a spring, stream or river. 4 Unstable ground is running silt, running sand, loose ground and the like.
	m³	2 Breaking up.	1 Rock. 2 Reinforced concrete. 3 Concrete. 4 Masonry or stonework.		1 Rock is any hard material which is of such size or location that it can only be removed by the use of wedges, rock hammers, special plant or explosives. 2 A boulder ≤ 5m³ in volume or one that can be lifted out in the bucket of an excavator will not constitute rock. 3 Degraded or friable rock that can be scraped out by the excavator bucket does not constitute rock. 4 Trimming exposed faces of rock is deemed included. 5 Breaking up hard surface pavings is measured in accordance with rule 5.3 of this section.

EFFECTIVE FROM 1 JANUARY 2013 RICS NEW RULES OF MEASUREMENT | 135

Item or work to be measured	Unit	Level one	Level two	Level three	Notes, comments and glossary
7 Extra over all types of excavation irrespective of depth—cont.	m	3 Excavating alongside existing underground services.		1 Nature of precaution required.	1 These items are measured where there is a risk of the existing service being affected by the excavation process. The method of protection is left to the discretion of the contractor. 2 If in doubt the surveyor must measure an item giving the nature of the live service.
	nr	4 Excavating across existing underground services			
8 Support to face(s) of excavation where not at the discretion of the contractor	m²	1 Maximum depth stated.	1 Location stated.	1 Method of forming support where not left to discretion of the contractor.	1 This work shall only be measured where it has been specifically specified in the contract documents or if the contractor has been instructed by the contract administrator to provide the support during the course of the works.
9 Disposal	item	1 Ground water.	1 Depth below original ground level stated. 2 Polluted water described if known.	1 If the post-contract water level differs from the pre-contract level the measurements must be revised accordingly.	1 The method and place of discharge left to the discretion of the contractor unless stated otherwise.
		2 Excavated material off site.	1 Destination if not at the contractor's discretion. 2 Hazardous material. 3 Non-hazardous material where it requires to be disposed to a specific location.		1 Irrespective of where excavated material originates.
10 Retaining excavated material on site	m³	1 Top soil. 2 All other excavated material.	1 To temporary spoil heaps. 2 Average distance to spoil heap stated.	1 Specified handling; details stated.	1 If the distance to spoil heap is not stated the location is left to the discretion of the contractor.

Item or work to be measured	Unit	Level one	Level two	Level three	Notes, comments and glossary
11 Filling obtained from excavated material	m²	1 Final thickness of filling not exceeding 500mm deep, finished thickness stated.	1 Source, distance, destination and method stated. 2 Maximum or average depth of layers stated.	1 Treatment of material prior to depositing in final location.	1 This includes topsoil and any other material arising from the excavations that have been specified to remain on site. 2 The thickness stated will be that after compaction. 3 Destinations will comprise general areas to make up levels, backfilling foundations, landscaping areas, planter beds and the like. 4 Compacting layers and surfaces are deemed included irrespective of depth and number of layers. 5 Source will be either direct from excavations or from temporary spoil heaps.
	m³	2 Final thickness of filling exceeding 500mm deep.			
12 Imported filling	m³	1 Blinding bed not exceeding 50mm thick, finished thickness stated.	1 Level, to falls, cross falls or cambers, stated. 2 Sloping not exceeding 15° from horizontal. 3 Sloping over 15° from horizontal.	1 Destination stated. Maximum or average depth of layers stated.	1 All types of surface treatments are deemed included. 2 The thickness stated will be that after compaction. 3 Compacting layers and surfaces are deemed included irrespective of depth and number of layers.
	m³	2 Beds over 50mm thick but not exceeding 500mm deep, finished thickness stated.			
	m³	3 Beds exceeding 500mm deep.			

Item or work to be measured	Unit	Level one	Level two	Level three	Notes, comments and glossary
13 Geotextile fabric 14 Radon barrier 15 Methane barrier 16 Damp proof membrane 17 Ground movement protection boards 18 Any other fabric, membrane or board: type stated 19 Ground stabilisation meshes and the like, type stated	m — — — — m²	1 Not exceeding 500mm wide, thickness or gauge stated. — — — — — — — — — 2 Over 500mm wide, thickness or gauge stated.	1 Horizontal. 2 Sloping. 3 Vertical.	1 Protective fleeces or boards, type stated. 2 Method of anchoring stated.	1 All turn-ups, turndowns, laps and joints deemed included. 2 Forming holes deemed included.
20 Cutting off tops of piles irrespective of length	nr	1 Size stated.			1 Cutting off tops of piles is deemed to include preparation and integration of reinforcement into pile cap or ground beam and disposal of all debris.

6   Ground remediation and soil stabilisation

		Notes, comments and glossary
Drawings that must accompany this section of measurement.	Mandatory information to be provided.	
	1 Site plan(s) showing location of work(s).	1 Ground conditions. 2 Ground water level(s) and date(s) established. 3 Nature of any known non-hazardous or hazardous contamination on the site or in the ground. 4 Starting level of each type of excavation.	
Minimum information that must be shown on the drawings or any other documents that accompany this section of measurement.	Works and materials deemed included.	
	1 Original and proposed ground levels. 2 Any item(s) that must remain on site during the works and be protected from damage. 3 Any item(s) adjacent to site that may impact on the works. 4 Details of trial pits or boreholes including their location. 5 Details of live over or underground services including their location. 6 Pile sizes and their locations where applicable.	1 Disposal off site of all surplus excavated material. 2 Disposal of all surface water. 3 Support to faces of excavation unless not at the discretion of the contractor. 4 Working space. 5 Excavation and filling for temporary works unless not at the discretion of the contractor. 6 Levelling, grading, trimming and compacting surfaces exposed by the excavations. 7 Curved work. 8 Multiple handling of excavated materials on site unless specified.	1 The ground water level must be re-established at the time each excavation is carried out and is defined as the post-contract ground water level. 2 Ground water levels subject to periodic changes due to tidal or similar effects are to be so described giving the mean high and low water levels.

Item or work to be measured	Unit	Level one	Level two	Level three	Notes, comments and glossary
1 Site dewatering	item	1 Area of site to be dewatered. 2 Maximum depth of boreholes.	1 Method of disposal of water stated if not at discretion of contractor.	1 Pre-contract water level. 2 Level to which ground water must be lowered to and maintained at.	1 Each type of remedial work shall be accompanied by a full description of the proposed works including limits on the extent, the proximity of adjoining building, restrictions on method, sequence and timing.
2 Sterilisation	m³	1 Maximum depth of ground to be treated.	1 Method of sterilisation stated.		
3 Chemical neutralising	m³	1 Maximum depth of ground to be treated.	1 Method of neutralisation stated.		
4 Freezing	m³	1 Maximum depth of ground to be treated.	1 Method of freezing stated. 2 Duration of freezing stated if not left to discretion of contractor.	1 Duration may be stated as a period of time or to a point within the contract programme such as 'completion of foundation work'.	
5 Ground gas venting	m²	1 Type of gas to be vented.	1 Method of collection stated. 2 Method of disposal stated.		
6 Soil nailing	m²	1 Area of site to be treated.	1 Length and diameter of nails, details stated. 2 Spacing of nails. 3 Method of grouting.		
7 Ground anchors	nr	1 Diameter and length of borehole.	1 Details stated.		

Item or work to be measured	Unit	Level one	Level two	Level three	Notes, comments and glossary
8 Pressure grouting/ground permeation	m²	1 Area and depth of site to be treated.	1 Details stated.		
9 Compacting	m²	1 Area of site to be treated 2 Weight of compactor	1 Details stated.		
10 Stabilising soil in situ by incorporating cement with a rotavator	m²	1 Area of site to be treated 2 Grams/m² of cement.	1 Details stated.		

7 Piling

Bored piling
Driven piling
Interlocking piling
Vibro-compacted stone piling

			Notes, comments and glossary
Drawings that must accompany this section of measurement.	1 Site plan showing site boundary and any adjacent buildings or features that might affect or be affected by the piling. 2 General piling layout.	Mandatory information to be provided. 1 Types of piles. 2 Nature of the ground. 3 Ground water level(s) and date(s) established. 4 Commencing levels from which work is expected to begin. 5 Limitations on headroom. 6 Kind and quality of materials. 7 Types of tests. 8 Type of grout. 9 Details of compaction.	1 Irregular ground levels must be stated. 2 Features that might affect piling include rivers, canals, tidal waters, flood areas and the like sufficiently close to the works that their situation may affect the piling process.
Minimum information that must be shown on the drawings that accompany this section of measurement.	1 Positions of piles. 2 Positions of existing buildings adjacent to the site. 3 Position of existing services.	Works and materials deemed included. 1 Temporary containment of spoil. 2 Any concrete placed in excess of the designed completed length. 3 Backfilling empty bores. 4 All pre-boring. 5 Re-positioning piling plant during the works 6 Maintaining all piling plant.	

Item or work to be measured	Unit	Level one	Level two	Level three	Notes, comments and glossary
1 Interlocking sheet piles	m²	1 Total driven area, maximum length stated.	1 Section modulus and cross section size, or section reference stated.	1 Removal to be stated if not left to discretion of contractor.	1 The cost of extraction is deemed included.
2 Bored piles 3 Driven piles 4 Other, type stated	m	1 Type stated. 2 Nominal size or diameter stated. 3 Total bored or driven length: maximum length stated. 4 Total concreted length. 5 Total number.	1 Contiguous/secant piling shall be so described. 2 Permanent casings shall be stated. 3 Raking, inclination stated.	1 Reinforcement to precast concrete piles.	1 Lengths are measured along the axes of the piles from the commencing level to the bottom of the pile.
5 Vibro-compacted piles	m	1 Nominal diameter. 2 Total number.			
6 Vibro-compacted trench fill	m	1 Nominal width and depth stated.			
7 Extra over piling	nr	1 Enlarged bases. 2 Enlarged heads.	1 Type of piling stated. 2 Diameter or size of enlargement stated.		1 This work is deemed to include everything necessary to form the enlargement including disposal of additional spoil.
	m	3 Pile extensions.	1 Total number stated. 2 Total concreted length.		1 Preparing head of pile to receive extension is deemed included.
8 Reinforcement to in-situ concrete piles	t	1 Nominal size and type of bars stated.			1 Reinforcement is deemed to include tying wire and spacers plus links and binders that are incorporated at the discretion of the contractor. 2 Types of bars include plain, deformed and helical.
9 Breaking through obstructions	hr	1 Rig standing.			1 Only measured where the obstruction is encountered above the founding stratum of the pile.

Item or work to be measured	Unit	Level one	Level two	Level three	Notes, comments and glossary
10 Disposal of excavated materials	m³	1 Off site. 2 On site.	1 Hazardous material. 2 Non-hazardous material where it requires to be disposed to a specific location. 3 Destination of spoil if not at the contractor's discretion.		1 The volume calculated is the nominal cross section by the pile lengths. 2 The volume of enlarged heads and bases is to be included.
11 Delays	hr	1 Rig standing.			1 Only measured where specifically instructed. 2 Deemed to include all associated labour, plant and overheads.
12 Tests	nr	1 Details stated.	1 Timing stated.		

8 Underpinning

		Mandatory information to be provided.	Notes, comments and glossary
Drawings that must accompany this section of measurement.	1 Location drawings. 2 Detailed section(s).	1 Limit of length of work to be carried out in one operation. 2 Maximum number of sections to be carried out at any one time. 3 Ground conditions including anticipated stability of excavations. 4 Ground water level(s) and date(s) established. 5 Nature of any known hazardous contamination on the site or in the ground. 6 Starting level of each type of excavation.	
		Works and materials deemed included.	1 If the work of underpinning is extensive the various elements of this work may be measured separately in accordance with the rules of the relevant trades or Work Sections, stating that the work is 'in underpinning'.
Minimum information that must be shown on the drawings that accompany this section of measurement.	1 Extent and method of work. 2 Details of existing structure to be underpinned.	1 Temporary support of existing structures. 2 Excavation and disposal. 3 Earthwork support. 4 Preliminary trenching. 5 All working space. 6 Disposal of ground and surface water. 7 Cutting away existing foundations and footings and disposal. 8 Preparing the underside of existing work. 9 Backfilling. 10 Surface treatments. 11 All new work associated with the process of underpinning. 12 All making good if specified.	

Item or work to be measured	Unit	Level one	Level two	Level three	Notes, comments and glossary
1 Underpinning	m m nr	1 Foundations. 2 Walls. 3 Bases.	1 Brief description of work stating depth, maximum width and method of underpinning. 2 Curved work.	1 Method, if not left to discretion of contractor, may be given by reference to drawing(s).	
		In the event that the underpinning is of such an extent that it can not easily be measured in accordance with these rules the works shall be measured in detail in accordance with the rules for the relevant trades required. In this case the work shall be described as 'In works of underpinning'.			1 Measured in accordance with the rules of the appropriate trade or Work Section stating that the work is 'in underpinning'.
2 Concrete	m³				
3 Formwork	m²				
4 Reinforcement	t				
5 Brickwork or blockwork	m²				
6 Tanking	m²				

9 Diaphragm walls and embedded retaining walls

Drawings that must accompany this section of measurement.	Mandatory information to be provided.		Notes, comments and glossary
1 Location drawings. 2 Site plans showing site boundary and any adjacent buildings or features that might affect or be affected by the construction of the diaphragm walls.	1 Limit of length of work to be carried out in one operation. 2 Maximum number of sections to be carried out at any one time. 3 Ground conditions including anticipated stability of excavations 4 Ground water level(s) and date(s) established. 5 Nature of any known hazardous contamination on the site or in the ground. 6 Starting level of each excavation.		1 Irregular ground shall be so described.

Minimum information that must be shown on the drawings that accompany this section of measurement.		Works and materials deemed included.	Level three
1 Extent of work.		1 Works and materials deemed included.	1 Excavation and disposal. 2 Earthwork support. 3 Preliminary trenching. 4 All working space. 5 Disposal of ground and surface water. 6 Backfilling. 7 Surface treatments.

Item or work to be measured	Unit	Level one	Level two	Level three	Notes, comments and glossary
1 Walls, thickness stated	m²	1 Commencing levels of excavation. 2 Maximum depth of excavation. 3 Finished top level of concrete if different from commencing level of excavation.	1 Details and method of construction stated. 2 Type of concrete. 3 Details of reinforcement.	1 Details of support fluid.	

Item or work to be measured	Unit	Level one	Level two	Level three	Notes, comments and glossary
2 Extra over excavation and disposal	m³	1 Breaking out hard materials. 2 Excavating in hazardous material.			
	m²	3 Breaking up hard surface pavings, thickness stated.			
3 Joints	m	1 Dimensioned description.	1 Method of forming.	1 Vertical. 2 Horizontal. 3 Raking. 4 Curved: radius stated.	
4 Trimming and cleaning exposed faces	m²	1 Details stated.			
5 Delays	hr	1 Authorised standing time.			1 Only measured where specifically instructed. 2 Deemed to include all associated labour, plant and overheads.
6 Tests	item	1 Details stated.			

10 Crib walls, gabions and reinforced earth

					Mandatory information to be provided.	Notes, comments and glossary
Drawings that must accompany this section of measurement.					1 Plans showing scope and location of each type of work.	
Minimum information that must be shown on the drawings that accompany this section of measurement.		1 Original ground levels. 2 Finished ground levels. 3 Condition of ground.		1 Kind and quality of materials.	1 Works and materials deemed included.	1 Final excavation means any minor trimming of earth surfaces required during each installation. All reduced level and foundation excavations, foundations and backfill are measured in accordance with the rules of Work Section 5: Excavating and filling.
					1 Final excavation associated with each installation. 2 Disposal of any excavated material including hazardous material. 3 Earthwork support. 4 Preparing surfaces to receive each installation. 5 Disposal of surface water.	

Item or work to be measured	Unit	Level one	Level two	Level three		Notes, comments and glossary
1 Crib walls	m²	1 Thickness stated.	1 Vertical. 2 Battering. 3 Curved on plan: radius stated.			1 Area is measured on front face. 2 No deductions made for voids ≤ 1.00m². 3 Dowels, pins, granular infill, compacting fill, special units, providing manufacturers certificates, building in pipes and forming openings ≤ 1.00m² deemed included.
2 Extra for	m	1 Ends. 2 Corners irrespective of angle.				
3 Gabion basket walls	m²	1 Basket size stated. 2 Gauge of basket wire stated. 3 Width of the base of the wall.	1 Vertical. 2 Battering: rate stated. 3 Sloping. 4 Curved on plan: radius stated. 5 Stepped on face.	1 Type of fill to baskets described. 2 Specified treatment of basket fill described. 3 Specified handling details stated.		1 Area is measured on front face. 2 Gabion baskets are deemed to include assembling, tying, fixing, bracing and tying lids. 3 Filling is deemed to include compaction and overfilling of fill material.

Item or work to be measured	Unit	Level one	Level two	Level three	Notes, comments and glossary
4 Earth reinforcement	m²	1 Mesh. 2 Fabric.	1 Horizontal. 2 Sloping. 3 Curved on plan: radius stated.	1 Method of anchoring stated. 2 Minimum laps stated. 3 Final facing, details stated.	1 Area measured in contact with base and excludes laps. 2 Assembling, tying, fixing, stacking and tensioning deemed included. 3 No deductions made for voids ≤ 1.00m².

11 In-situ concrete works

In-situ concrete
Surface finishes to in-situ concrete
Formwork
Reinforcement
Designed joints in in-situ concrete
Accessories cast in to in-situ concrete
In-situ concrete sundries

		Mandatory information to be provided.	Notes, comments and glossary
Drawings that must accompany this section of measurement.	1 General arrangement drawings.	1 Kind, quality and size of materials. 2 Details of tests of materials. 3 Details of tests of finished work. 4 Limitations on method, sequence, speed or size of pouring. 5 Method of compaction. 6 Method of curing. 7 Details of watertightness.	1 Work in substructures, superstructures or external works to be stated in headings or descriptions. 2 Watertight work shall be so described.
		Works and materials deemed included.	
Minimum information that must be shown on the drawings that accompany this section of measurement.	1 Relative position of all members. 2 The size of members. 3 The thickness of slabs. 4 The permissible loads in relation to casting times.	1 Concrete volume is measured net. 2 No allowance in volume to be made for deflection of formwork. 3 Deductions are not made for reinforcement, steel sections, cast-in accessories, voids ≤ 0.05m³ except voids in troughed and coffered slabs. 4 Concrete is deemed cast into formwork unless otherwise described. 5 Concrete is deemed finished as struck from basic finish formwork. 6 All top surfaces and soffits are deemed to finish horizontal unless otherwise stated. 7 All top surfaces are deemed finished tamped.	1 This applies to concrete laid on ribbed metal decking as well as other types of formwork.

Plain in-situ concrete
Reinforced in-situ concrete
Fibre reinforced in-situ concrete
Sprayed in-situ concrete

Item or work to be measured	Unit	Level one	Level two	Level three	Notes, comments and glossary
1 Mass concrete	m³	1 Any thickness.	1 In filling voids. 2 In trench filling. 3 In any other situation: details stated.	1 Poured on or against earth or unblinded hardcore.	1 Mass concrete is any unreinforced bulk concrete not measured elsewhere. 2 The volumes of each type of mass concrete work may be aggregated or given separately.
2 Horizontal work	m³	1 ≤ 300 thick. 2 > 300 thick.	1 In blinding. 2 In structures.	1 Poured on or against earth or unblinded hardcore. 2 Reinforced > 5%.	1 Horizontal work includes blinding, beds, foundations, pile caps, column bases, ground beams, slabs, coffered and troughed slabs, landings, beams, attached beams, beam casings, shear heads, upstands whose height is ≤ than three times their width, kerbs, copings. 2 The volumes of each type of horizontal work may be aggregated or given separately. 3 Work laid in bays shall be so described giving average area of bays.
3 Sloping work ≤ 15° 4 Sloping work > 15°	m³	1 ≤ 300 thick. 2 > 300 thick.	1 In blinding. 2 In structures. 3 In staircases.	1 Poured on or against earth or unblinded hardcore. 2 Reinforced > 5%.	1 Sloping work includes blinding, beds, slabs, steps and staircases, kerbs, copings. 2 Includes any attached beams, upstands, shear heads or similar. 3 The volumes of each type of sloping work may be aggregated or given separately. 4 Work laid in bays shall be so described giving average area of bays.

EFFECTIVE FROM 1 JANUARY 2013

Item or work to be measured	Unit	Level one	Level two	Level three	Notes, comments and glossary
5 Vertical work	m³	1 ≤ 300 thick. 2 > 300 thick.	1 In structures.	2 Reinforced > 5%.	1 Vertical work includes columns, attached columns, column casings, walls, retaining walls, filling to hollow walls, parapets or upstand beams where height is greater than three times the width. 2 The volumes of each type of vertical work may be aggregated or given separately.
6 Sundry in-situ concrete work	m m³	1 Work ≤ 300 wide or thick. 2 Work > 300 wide or thick.	1 Horizontal. 2 Sloping. 3 Vertical.	4 Reinforced > 5%.	1 Includes work such as backsills, machine and plant bases and the like.
7 Sprayed in-situ concrete	m²	1 Thickness stated.	1 Slabs.	1 Tops. 2 Soffits. 3 Curved.	1 The method of application and finish to be stated in the description.
			2 Walls. 3 Beams. 4 Columns.	1 Curved.	
Surface finishes to in-situ concrete					
8 Trowelling 9 Power floating 10 Hacking 11 Grinding 12 Any other surface treatment not left to discretion of the contractor	m²	1 To top surfaces. 2 To faces. 3 To soffits.	1 Sloping. 2 Falls. 3 Crossfalls.	1 Application of surface hardeners, sealers, dust proofers, waterproofers, carborundum grains or the like shall be so described.	

Item or work to be measured	Unit	Level one	Level two	Level three	Notes, comments and glossary
Formwork		1 Plain formwork. 2 Special finish formwork.			1 Plain finish shall be left to discretion of the contractor. 2 Special finishes shall be described. 3 Curved work shall be described stating the radii. 4 Permanent formwork or formwork left in shall be so described. 5 Void formers shall be so described. 6 No deductions shall be made for voids ≤ 5.00m² 7 All kickers except to walls shall be deemed included. 8 Top formwork is measured for sloping surfaces that are > 15° or where otherwise specifically required. 9 All square, raking and curved cutting deemed included. 10 All holes, boxings, recesses, rebates, chamfers, nibs, channels and the like are deemed included.
13 Sides of foundations and bases 14 Edges of horizontal work	m m²	1 ≤ 500 high: width stated. 2 > 500 high.			
15 Soffits of horizontal work 16 Soffits of troughed or waffled horizontal work, details described	m²	1 For concrete ≤ 300 thick. 2 For concrete 300 to 450 thick. 3 For concrete > 450 thick.	1 Propping ≤ 3m high. 2 Propping over 3m but not exceeding 4.5m high. 3 And thereafter in 1.5m stages.		1 Includes suspended slabs and stair landings. 2 Through propping to be described if not left to discretion of contractor.
17 Sides and soffits of isolated beams 18 Sides and soffits of attached beams 19 Sides of upstand beams 20 Sides of isolated columns, nr stated 21 Sides of attached columns	m²	1 Regular: shape stated. 2 Irregular shaped, dimensioned description or diagram.			1 Shape is deemed regular unless described as otherwise. 2 Irregular shape is any shape other than square or rectangular. 3 Includes concrete casings to steel beams and columns.

Item or work to be measured	Unit	Level one	Level two	Level three	Notes, comments and glossary
22 Faces of walls and other vertical work	m²	1 Vertical. 2 Battered one face. 3 Battered both faces.	1 Rate of batter to be stated.		1 Work to single sides shall be so described.
23 Extra over	nr	1 Openings for doors or the like: thickness of wall stated.	1 ≤ 5.00m². 2 5.00m² to 10.00m². 3 > 10.00m².		1 All additional labour and material needed to form the opening is deemed included.
24 Wall ends, soffits and steps in walls	m m²	1 ≤ 500 wide, width stated. 2 > 500 wide.			1 Excludes ends and soffits of walls created by the formation of an opening. These are deemed included in the item for forming the opening.
25 Soffits of sloping work	m²	1 Sloping one way. 2 Sloping two ways.			1 This includes work to soffits of slabs, ramps, steps, staircases and the like.
26 Staircase strings and the like	m	1 Maximum width stated.			
27 Staircase risers and the like	m	1 Vertical: width stated. 2 Undercut: width stated.			
28 Sloping top surfaces	m²	1 ≤ 15°. 2 > 15°.			
29 Steps in top surfaces 30 Steps in soffits	m	1 ≤ 500 high: width stated. 2 > 500 high.			
31 Complex shapes	nr	1 Dimensioned description or diagram.	1 Propping ≤ 3m high. 2 Propping over 3m but not exceeding 4.5m high. 3 And thereafter in 1.5m stages.		
32 Wall kickers	m	1 Plain. 2 Suspended.			1 Length is measured along centre line and is deemed to include both sides.

Item or work to be measured	Unit	Level one	Level two	Level three	Notes, comments and glossary
Reinforcement					
33 Mild steel bars 34 High yield steel bars	t	1 Nominal size stated.	1 Straight. 2 Bent. 3 Curved. 4 Links.	1 Bars exceeding 12m long: length stated. 2 Deformed. 3 Bending restrictions	1 Forming hooks, tying wire, spacers, cutting, and bending is deemed included. 2 Chairs and connectors are deemed included unless not at discretion of contractor.
35 Accessories not at the discretion of the contractor	nr	1 Nominal size stated.	1 Chairs or stools. 2 Connectors.		
36 Pre/Post-tensioned members	nr	1 Dimensioned description. 2 Nominal size stated. 3 Method of tensioning stated.	1 Composite construction described.	1 Sleeves, tendons, fittings and grouting described.	1 Post-tensioning is measured by the number of tendons in identical members.
37 Mesh	m²	1 Weight per m² stated. 2 Fabric reference stated 3 Minimum laps stated.	1 Bent. 2 Strips in one width, width stated.		1 Laps, tying wires, all cutting, bending, spacers, stools, chairs and other supports deemed included. 2 Voids ≤ 1.00m² in area not deducted. 3 Bent fabric is deemed to include fabric that is wrapped around steel members.

Item or work to be measured	Unit	Level one	Level two	Level three	Notes, comments and glossary
Designed joints in in-situ concrete					1 Joints located at the discretion of the contractor are not measured. 2 Details of primers, cleaners, fillers, waterstops, backing strips, reinforcement, ties, sealants, the method of application, preparation and the like shall be stated in the description.
38 Plain 39 Formed 40 Cut	m	1 Dimensioned description. 2 total depth stated.	1 Horizontal. 2 Vertical. 3 Curved, radius stated.		1 Plain joints are those that do not require formwork. 2 Formed joints are deemed to include formwork. 3 All preparation, cleaners, primers and sealers are deemed included. 4 All angles, ends, intersections are deemed included whether they are formed, welded or purpose made.
Accessories cast into in-situ concrete					1 Kind, quality of materials and size or manufacturers reference shall be stated.
41 Type or proprietary reference stated	m² - - - - m - - - - nr	1 Dimensioned description.	1 If linear or superficial quantities are used the description must include any appropriate spacing dimensions.		1 Cast-in accessories include anchor bolts, anchor boxes, fixing bolts, dowels, column guards, isolated glass blocks and any other ancillary item that is specified to be cast in as the concrete work proceeds. 2 Cast-in accessories exclude reinforcement, tying wire, distance blocks, spacers, chairs, structural steel members, hollow blocks, filler blocks, void formers, permanent formwork, joints, all components around which concrete is cast but which are not fixed in position by the contractor.

In-situ concrete sundries

Item or work to be measured	Unit	Level one	Level two	Level three	Notes, comments and glossary
42 Grouting	nr	1 Dimensioned description.	1 Stanchion bases. 2 Grillages.		1 Formwork or other temporary means of support to exposed edges and the like is deemed included.
43 Filling mortices or holes	nr				
44 Filling chases	m				

12 Precast/composite concrete

Precast/composite concrete walls, partitions and panels
Precast/composite concrete decking and flooring
Other precast/composite concrete work

		Mandatory information to be provided.	Notes, comments and glossary
Drawings that must accompany this section of measurement.	1 General arrangement drawings. 2 Specific drawings or details relating to precast works.	1 Kind and quality of materials. 2 Sizes and spacing of planks and blocks. 3 Kind, quality and mix of concrete. 4 Methods of compaction and curing. 5 Bedding and fixing. 6 Surface finishes. 7 Kind and quality of reinforcement or pre/post-tensioning, spacing and stresses. 8 Finish of exposed surfaces. 9 Details of tests of materials. 10 Details of tests of finished work.	
		Works and materials deemed included.	
Minimum information that must be shown on the drawings that accompany this section of measurement.	1 The relative position of concrete members. 2 Size of members. 3 Thickness of slabs and panels. 4 Permissible loads.	1 Moulds and formwork. 2 Reinforcement. 3 Bedding. 4 Fixings. 5 Temporary support. 6 Cast-in accessories. 7 Pre-tensioning or pre-stressing. 8 Filled ends. 9 All grouting. 10 Margins ≤ 500mm wide.	

Item or work to be measured	Unit	Level one	Level two	Level three	Notes, comments and glossary
1 Composite concrete work	m²	1 Nature of work described. 2 Composition of work described. 3 Thickness stated.	1 Horizontal. 2 Sloping ≤ 15°. 3 Sloping > 15°. 4 Vertical.	1 Reinforcement: details stated. 2 Post-tensioning: number of tendons and details stated. 3 Cast-in accessories: details stated.	1 Will apply to panels, slabs, walls, partitions, decking. 2 The thickness stated is the combined thickness of both precast and in-situ work. 3 Margins greater than 500mm wide are measured as ordinary in-situ concrete slabs. 4 No deduction is made for voids ≤ 1.00m².
2 Designed joints	m	1 Dimensioned description.	1 Type and sizes of filling and sealant stated.		1 Joints formed at the discretion of the contractor are not measured.
3 Holding down or tie straps	nr	1 Dimensioned description or proprietary reference number.	1 Material type stated. 2 Protective coating stated.	1 Method of fixing.	

13 Precast concrete

Precast concrete frame structures

Precast concrete sills, lintels, copings and other units

Precast concrete panel cladding

Precast concrete slabs

Precast concrete walls and partitions

Precast concrete decking

Precast concrete rooflights and pavement lights

Precast composite concrete work

Drawings that must accompany this section of measurement.	Mandatory information to be provided.	Notes, comments and glossary
1 General arrangement drawings. 2 Specific drawings or details relating to precast works.	1 Kind, quality and mix of concrete. 2 Methods of compaction and curing. 3 Bedding and fixing. 4 Surface finishes. 5 Kind and quality of reinforcement or pre/post-tensioning, spacing and stresses. 6 Finish of exposed surfaces. 7 Details of tests of materials. 8 Details of tests of finished work.	

Minimum information that must be shown on the drawings that accompany this section of measurement.			Works and materials deemed included.		
			1 The relative position of concrete members. 2 The thickness or size of members. 3 Thickness of slabs and panels. 4 Permissible loads. 5 Full details of anchorages, ducts, sheathing and vents.	1 Moulds and formwork. 2 Reinforcement. 3 Bedding. 4 Fixings. 5 Temporary support. 6 Cast-in accessories. 7 Pre-tensioning or pre-stressing. 8 Filled ends. 9 All grouting. 10 Margins. 11 Angles and fair ends. 12 Glass lenses.	

Item or work to be measured	Unit	Level one	Level two	Level three	Notes, comments and glossary
1 Precast concrete goods	nr	1 Dimensioned description or dimensioned diagram.	1 Reinforcement: details stated. 2 Post-tensioning; number of tendons and details stated. 3 Cast-in accessories: details stated.	1 Stoolings; number stated.	1 Will apply to sills, lintels, padstones, staircases, landings, panels, partitions, columns, beams, structural frames and other precast features.
	m	2 Dimensioned description stating number of pieces.	4 Sloping not exceeding 15°. 5 Sloping exceeding 15°.		2 Will apply to frame members, copings and the like.
	m²	3 Dimensioned description stating thickness.		2 Span where relevant.	3 Will apply to panels, slabs, walls, partitions, decking and the like.
2 Rooflights 3 Pavement lights 4 Vertical panel lights	m² ----- nr	1 Dimensioned description. Number stated. ------------------ 2 Dimensioned description.	1 Sizes and extent of reinforcement stated.		1 Isolated glass lenses are measured as accessories cast into in situ concrete. 2 Roof, pavement and panel lights are deemed to include moulds, formwork, temporary propping, reinforcement, bedding and glass lenses.

Item or work to be measured	Unit	Level one	Level two	Level three	Notes, comments and glossary
5 Designed joints	m	1 Dimensioned description.	1 Type and sizes of filling and sealant stated.		1 Joints formed at the discretion of the contractor are not measured.
6 Holding down or tie straps	nr	1 Dimensioned description or proprietary reference number.	1 Material type stated. 2 Protective coating stated.	1 Method of fixing.	

14 Masonry

Brick/block walling
Glass block walling
Natural stone rubble walling and dressings
Natural stone ashlar walling and dressings
Artificial/cast stone walling and dressings

		Mandatory information to be provided.	Notes, comments and glossary
Drawings that must accompany this section of measurement.	1 Plans of each floor level. 2 Principal sections showing floor to floor heights. 3 External elevations. 4 Any other major masonry work.	1 Kind, quality and size of brick/block/stone units. 2 Type of finish/facings to each side. 3 Bond. 4 Composition and mix of mortar. 5 Type of pointing. 6 Bonding to other work. 7 Raking or curved work so described. 8 Radius of curved work.	1 All walling is measured on the centre line irrespective of construction. 2 Thicknesses stated are nominal. 3 Work is deemed vertical unless otherwise described. 4 All wall dimensions exclude applied finishes. 5 No deductions will be made for voids or built in items whose cross sectional area is less than 0.50m².
Minimum information that must be shown on the drawings that accompany this section of measurement.	1 Major horizontal and vertical dimensions. 2 Types of materials used to construct the walls and other structures. 3 Position of main structural frame members.	Works and materials deemed included. 1 All rough and fair cutting. 2 All ends and angles, either formed or proprietary. 3 Extra material for curved work. 4 Forming all rough and fair grooves, throats, mortices, chases, rebates, holes, stops, mitres and the like labours. 5 Raking out joints to form a key. 6 Raking out joints to insert flashings. 7 Centering or other forms of temporary support.	

Item or work to be measured	Unit	Level one	Level two	Level three	Notes, comments and glossary
				8 Labour in eaves filling. 9 Labours in returns, ends and angles. 10 Centering. 11 Overhand work. 12 Bonding ends of walls to other work. 13 All extra material required for bonding. 14 Additional material in laps. 15 Preparation of all surfaces to receive subsequent applications.	
1 Walls; overall thickness stated 2 Diaphragm walls; overall thickness stated, spacing and thickness of ribs stated.	m²	1 Brickwork. 2 Blockwork. 3 Glass blockwork. 4 Natural stone. 5 Cast stone. 6 Other: type stated.	1 Skins of hollow walls. 2 Battered. 3 Tapered; one side or both sides. 4 Built against other work. 5 Used as formwork.	1 Method of forming.	1 The description shall describe the type of construction of the masonry such as rubble or ashlar work and height of coursing. 2 Walls are measured on the centre line of the material unless otherwise stated. 3 The radius of curved work is taken from the centre line. 4 Battering walls are sloping walls with parallel sides. 5 Thickness stated for tapering walls is the mean thickness. 6 No deductions to be made for voids or built in items whose cross sectional area equal to or less than 0.50m². 7 Thickness stated for diaphragm walls is total thickness of both skins and cavity void.
3 Vaulting; thickness and type stated	m²				

Item or work to be measured	Unit	Level one	Level two	Level three	Notes, comments and glossary
4 Isolated piers; isolated casings; chimney stacks; columns	m	1 Dimensioned description or dimensioned diagram.	1 Vertical. 2 Battered. 3 Curved: radius stated.		1 These are such when their length on plan is equal to or less than four times their thickness except where caused by openings. 2 No deductions to be made for flues and the like equal to or less than 0.10m² in cross sectional area.
5 Attached projections	m		1 Vertical. 2 Raking. 3 Horizontal. 4 Curved: radius stated.		1 Attached projections are attached piers whose length on plan is equal to or less than four times their thickness plus plinths, oversailing courses and the like.
6 Arches (number stated)	m	1 Height of face, width of soffit and shape of arch stated.			1 The length is the mean girth or length on face.
7 Bands; dimensioned description	m	1 Flush. 2 Sunk: depth of set back stated. 3 Projecting: depth of set forward stated.	1 Vertical. 2 Raking. 3 Horizontal. 4 Curved: mean radius on face stated.	1 Entirely of stretchers. 2 Entirely of headers. 3 Alternate headers/stretchers. 4 Other bond.	1 Bands are brick-on-edge bands, brick-on-end bands, basket pattern bands, moulded or splayed cappings, moulded string courses, moulded cornices and the like.
8 Flues 9 Flue linings 10 Filling around flues	m	1 Dimensioned description.	1 Method of forming.		
11 Extra over walls for perimeters and abutments, details stated	m	1 Dimensioned description.	1 Method of forming. 2 Closing cavities, additional ties, insulation and all other associated work is deemed included.		1 This will include work forming eaves, copings, kerbs, quoins, ends and the like.
12 Extra over walls for opening perimeters, details stated	m				1 This will include work forming sills, jambs, reveals, cavity closers, architraves, lintels, mullions, transoms, thresholds, steps and the like.

Item or work to be measured	Unit	Level one	Level two	Level three	Notes, comments and glossary
13 Special purpose blocks or stones	nr	1 Dimensioned description.	1 Function stated.		1 Descriptions of stones are given as the smallest block from which each item can be obtained having regard in the case of natural stone to the plane in which the stone is required to be laid with relation to its quarry bed. The dimensions are taken over one mortar bed and one mortar joint.
14 Forming cavity	m²	1 Width and method of forming.	1 Type and spacing of ties.		
15 Cavity insulation	m²	1 Type and thickness.	2 Method of installing or fixing.		
16 Damp-proof courses ≤ 300mm wide	m	1 Gauge or thickness.	1 Vertical.		1 Damp-proof courses are deemed to include: (a) forming laps, ends and angles. (b) pointing exposed edges. (c) bonding to damp-proof membranes and the like.
17 Damp-proof courses > 300mm wide	m²	2 Number of layers.	2 Raking.		
18 Pre-formed cavity trays	m	3 Composition and mix of bedding materials.	3 Horizontal. 4 Curved, mean radius on face stated. 5 Stepped.		
19 Joint reinforcement	m	1 Width stated.			
20 Fillets	m	1 Width and thickness stated.		1 Weather. 2 Angle. 3 Other.	
21 Pointing	m	1 Width and depth of joint to be pointed.		1 Type of finish.	1 This relates to the pointing of joints, flashings, frames and the like.
22 Joints	m	1 Width and thickness stated.		1 Type of filler, sealant, pointing, method of application and preparation. 2 Ties: centres stated. 3 Channels.	1 Joints are only measured where their composition and position is designed.

Item or work to be measured	Unit	Level one	Level two	Level three	Notes, comments and glossary
23 Wedging and pinning	m	1 Width and thickness stated.			
24 Creasing	m	1 Width stated. 2 Number of courses.			
25 Proprietary and individual spot items.	nr	1 Dimensioned description or dimensioned diagram.	1 Proprietary reference or catalogue number where applicable.	1 Method of forming, building in or fixing.	1 All of these items are deemed to include all necessary forming of openings or pockets, liners, cavity closers, damp-proof courses, fixings and fastenings and builder's work in connection with any associated mechanical or electrical connections. 2 This will include items such as windposts, head restraint channels, steel lintels, wall end ties, wall end bonding channels, cappings, chimney pots, finials, boiler seats, soot doors, plinths, steps, winders, landings, bases, key blocks, air bricks, ventilator gratings, flue blocks, fire backs and sides, grates, carved bricks, blocks or stones, quoin stones, jamb stones, hearths, weep-hole formers and the like. 3 The list is not exhaustive. Any type of individual shaped piece should be enumerated and described either by a dimensioned description or a dimensioned diagram or by reference to the specification or trade brochure.

15 Structural metalwork

Structural Steelwork
Structural aluminium work

		Mandatory information to be provided.	Notes, comments and glossary.
Drawings that must accompany this section of measurement.	1 All General arrangement drawings, Plans, Sections and Elevations. 2 Drawings specific to the work.		1 Types and grade of materials including steel and steel to steel bolts. 2 Specification describing fabrication, welding, testing, erection and everything else necessary to complete the installation. 3 Surface preparation prior to application of any surface treatment or finish.
Minimum information that must be shown on the drawings that accompany this section of measurement.	1 Position of the work in relation to the proposed structure and any existing structure that the new work is being connected to. 2 The types and sizes of all structural members and their positions in relation to each other. 3 Details of connections or of the reactions, moments and axial loads at connection points.	Works and materials deemed included.	1 The mass of framing includes all components. 2 No allowance is made for the mass of welds, bolts, nuts, washers, rivets, and protective coatings. 3 Permanent erection is deemed to include all specified operations subsequent to fabrication including delivery to site.

Item or work to be measured	Unit	Level one	Level two	Level three	Notes, comments and glossary
1 Framed members, framing and fabrication	t	1 Lengths not exceeding 1.00m 2 Lengths over 1.00 but not exceeding 9.00m 3 Lengths exceeding 9.00m	1 Weight ≤ 25kg/m. 2 Weight 25–50kg/m. 3 Weight 50–100kg/m. 4 And so on in increments of 50kg/m.	1 Columns. 2 Beams. 3 Rafters. 4 Bracings. 5 Purlins and cladding rails. 6 Grillages. 7 Trusses. 8 Plate girders. 9 Framing to doors and windows. 10 Trimmers to roofs and walls. 11 Rafter and column stays, number stated.	1 Castellated. 2 Tapered. 3 Curved. 4 Cambered. 5 Hollow: shape stated. 6 Built-up work: details of construction stated. 7 Compound fabrications, details of construction stated. 8 Cellular. 9 Secondary steelwork. 10 Temporary bracing or support work not at discretion of the contractor; subsequent removal is deemed included unless stated otherwise.
2 Framed members, permanent erection on site				1 Crane rails.	1 Details and centres of fixing clips and resilient pads stated.
				1 Wires, cables, rods, ties and bars.	
3 Isolated structural members, fabrication	t	1 Plain member: use stated.	1 Dimensioned description and weight per lin m		1 An isolated structural member is one not integrally connected to other structural members.
4 Isolated structural members, permanent erection on site	nr	2 Built-up member: use stated.			
5 Allowance for fittings	t	1 Calculated weight. 2 Percentage allowance; percentage stated.	1 To framed members. 2 To isolated members.		1 Fittings are components that allow members to be joined together or are other brackets, supports and the like that are supplied and attached, either on or off site, to the main loadbearing frame by the structural metal contractor.

Item or work to be measured	Unit	Level one	Level two	Level three	Notes, comments and glossary
6 Cold rolled purlins, cladding rails and the like	m	1 Type and method of fixing stated. 2 Size or proprietary reference stated.	1 Purlins and cladding rails. 2 Sag rods. 3 Stays. 4 Other.	1 Method of fixing stated.	
7 Extra over for	nr	1 Forming cranks.	Dimension of member stated.		
8 Profiled metal decking, type and/or profile stated	m²	1 Height not exceeding 5.00m 2 Height exceeding 5.00m but not exceeding 10.00m 3 And so on in 5.00m increments.	1 Shear studs: size and spacing stated.	1 Method of fixing.	1 This work is only measured here when it forms part of the structural steel package otherwise it would be measured in accordance with the rules for permanent formwork. 2 Heights are always measured from finished floor level unless otherwise stated. 3 The area of metal decking is calculated as the finished area of concrete cast onto the decking.
9 Extra over for	m	1 Edge trims; size, girth, profile or proprietary reference stated.	1 Straight. 2 Raking. 3 Curved, radius stated.		1 Ends, angles, intersections are all deemed included.
	m	2 Curved cutting.			
10 Holding down bolts or assemblies	nr	1 Type and diameter stated.	1 Anchor plates, frames, members, tubes, cones and any other associated accessory stated	1 Supply only.	
11 Special bolts	nr	1 Type and diameter stated.		1 Background if other than structural steel.	1 Special bolts are all bolts and fasteners other than grade 4.6 black bolts and holding down assemblies.

Item or work to be measured	Unit	Level one	Level two	Level three	Notes, comments and glossary
12 Connections to existing steel and other members or structures	nr	1 Details stated.			1 All labours on new and existing steel are deemed included. 2 Labours on non-steel structures are measured elsewhere.
13 Trial erection	t	1 Details and location stated.			1 Only measured when not at discretion of contractor. 2 The information given in the description must include items such as the number of erectable pieces and number of site welds. This information may be given by reference to drawing(s).
14 Filling hollow sections	item	1 Water. 2 Concrete. 3 Other material type stated	1 Details stated.		
15 Surface treatments	m²	1 Galvanising. 2 Sprayed coating. 3 Painting. 4 Other treatment: type stated.	1 On site. 2 Off site.	1 Preparation described. 2 Number of coats stated. 3 Thickness of coat(s) stated. 4 Fire rating stated. 5 Finish stated.	1 All preparation is deemed included.
16 Isolated protective coatings	nr	1 Approximate size or area stated.	2 Type of protective coated stated. 3 Preparation described.	1 On site. 2 Off site.	
17 Testing	item	1 Load tests. 2 Fire protection tests. 3 Other tests.	1 Details stated.		

16 Carpentry

Timber framing
Timber first fixings
Timber, metal and plastic boarding, sheeting, decking, casings and linings
Metal and plastic accessories

Drawings that must accompany this section of measurement.	Mandatory information to be provided.	Notes, comments and glossary
1 All general arrangement plans, sections and elevations. 2 Drawings specific to the work.	1 Kind, quality and size of materials. 2 Grade of timber. 3 Type of preservative treatment. 4 Type of protective coating 5 Method of fixing where not at the discretion of the contractor. 6 Method of jointing or construction where not at the discretion of the contractor. 7 Spacing of battens and grounds. 8 Nature of base.	
Minimum information that must be shown on the drawings that accompany this section of measurement.	Works and materials deemed included.	
1 Layout and spacing of timbers. 2 Types of materials used. 3 Position of timbers in relation adjacent structures.	1 All sizes nominal unless otherwise stated. 2 All timbers are sawn unless otherwise stated. 3 All work fabricated, assembled and erected on site unless otherwise stated. 4 All work fixed by nails unless otherwise stated. 5 Holes in timber for bolts and all other fixings. 6 All labours on timber. 7 All webs, gussets and the like on trusses and portals.	

Item or work to be measured	Unit	Level one	Level two	Level three	Notes, comments and glossary
1 Primary or structural timbers	m	1 Nominal size stated.	1 Rafters and associated roof timbers. 2 Purlins. 3 Wall plates. 4 Roof and floor joists. 5 Beams. 6 Posts or columns. 7 Partition and wall members. 8 Strutting.	1 Selection and protection for subsequent treatment. 2 Matching grains or colours. 3 Limits on planing margins. 4 Fixing centres stated where not at discretion of the contractor. 5 Type and size of bolts and other fixings and fastenings. 6 Background or nature of base. 7 Fixing through vulnerable materials. 8 Surface finish or treatment applied as part of the manufacturing process. 9 Length > 6.00 in one continuous length.	1 Strutting is measured through the structural members being stiffened.
2 Engineered or prefabricated members/items	nr	1 Finished Dimensioned description.	2 Roof trusses. 3 Portal frames. 4 Trusses. 5 Wall panels. 6 Beams. 7 Joists. 8 Posts or columns.	1 Manufacturers' reference. 2 Length > 6.00 in one continuous length.	
3 Backing and other first fix timbers	m m²	1 Nominal Dimensioned description of each member. 2 Nominal Dimensioned description of each member.	1 Grounds. 2 Battens. 1 Framed grounds, battens and bracketing.	1 Centres, each way stated.	
4 Boarding, flooring, sheeting, decking, casings, linings, sarking, fascias, bargeboards, soffits and the like	m m²	1 Not exceeding 600mm wide; finished width and thickness stated. 2 Over 600mm wide; finished thickness stated.	1 Horizontal. 2 Sloping. 3 Vertical. 4 Curved: radius stated. 5 Soffit or ceiling. 6 Other shape described.	1 Finish stated unless sawn. 2 Type of joints where not at the discretion of the contractor. 3 Profile.	1 The location of the work shall be given such as external walls, internal walls and attached piers, isolated columns, floors, ceilings and attached beams, isolated beams, roofs, eaves, verges, tops and cheeks of dormers and the like.

Item or work to be measured	Unit	Level one	Level two	Level three	Notes, comments and glossary
5 Ornamental ends of timber members	nr	1 Size and detail stated.			
6 Metal fixings, fastenings and fittings	nr	1 Dimensioned description. 2 Proprietary reference or catalogue number where applicable.	1 Ties. 2 Rods. 3 Brackets. 4 Straps. 5 Shoes. 6 Bolts. 7 Joist hangers. 8 Other: details stated.	1 Method of fixing stated. 2 Background stated if not timber	1 Bolts include heads, nuts and washers. 2 The length of a bolt is measured over the head. 3 Work is deemed to include all labours in fabricating and fixing including drilling holes in the fitting, the timber and the background.

17 Sheet roof coverings

Bituminous felts
Plastic sheets
Sheet metals
Rigid boards with factory applied sheet coverings
Any other type of sheet roof covering

Drawings that must accompany this section of measurement.	Mandatory information to be provided.	Notes, comments and glossary
1 Plans of each roof. 2 Principal sections of roof. 3 External elevations.	1 Kind, quality of materials. 2 Thickness, gauge, weight and temper. 3 Method of fixing. 4 Details and position of laps, drips, welts, beads, rolls, joints, upstands and downstands. 5 Bonding to other work. 6 Radius of curved work. 7 Type(s) and spacing of joints. 8 Type(s) and spacing of seams. 9 Special finishes.	1 Examples of roof coverings measured in accordance with the rules of this section are: sheet lead, aluminium, copper, zinc, stainless steel, fibre bitumen, butyl rubber, thermoplastic and the like membranes. 2 This list is not exhaustive.

Minimum information that must be shown on the drawings that accompany this section of measurement.			Works and materials deemed included.	The areas and lengths measured are net in contact with base.
			1 Major horizontal and vertical dimensions.	1 Underlay in contact with the covering.
			2 Types of materials used to construct the roof and its structures.	2 All rough and fair cutting and waste.
			3 Position of main structural frame members.	3 Extra material required for bonding.
			4 Height of work above ground level stating whether working platform level or finished level.	4 Extra material in laps and dressings.
			5 All labours and dressings.	5 Extra material for curved work.
				6 All ends, angles and intersections either formed or proprietary.
				7 Forming all rough and fair grooves, throats, mortices, chases, rebates, holes, stops, mitres and the like labours.
				8 Raking out joints to insert flashings.
				9 Labours in returns, ends and angles.
				10 Work in isolated areas.
				11 Work in forming voids and holes ≤ 1.00m².

Item or work to be measured	Unit	Level one	Level two	Level three	Notes, comments and glossary
1 Coverings > 500mm wide	m²	1 Horizontal. 2 Sloping; pitch stated. 3 Vertical. 4 Curved: radii stated.	1 Underlays. 2 Insulation. 3 Finish to exposed surface.	1 Nature of base.	1 No deduction is made for voids ≤ 1.00m². 2 Finishes include solar reflective paint, chippings and the like. They exclude tiles, paving slabs, grass roofs and the like.
2 Coverings ≤ 500mm wide	m				

Item or work to be measured	Unit	Level one	Level two	Level three	Notes, comments and glossary
3 Extra over for forming	m	1 Drips. 2 Welts. 3 Rolls. 4 Seams. 5 Laps.	1 Height stated. 1 Height stated. 1 Width and height stated. 1 Length stated. 1 Length stated.	1 Type stated. 1 Wood core. 2 Hollow.	1 Lengths shall be net length of each labour. 2 All additional sheet material required to form the labour item shall be deemed included. 3 All additional underlay, insulation and surface finish shall be deemed included.
4 Boundary work, location and method of forming described	m	1 Net girth stated. 2 Average net girth stated, number of lengths stated.	1 Abutments. 2 Eaves. 3 Ridges. 4 Verges. 5 Valleys. 6 Hips. 7 Vertical angles. 8 Upstands ≤ 500mm high. 9 Downstands ≤ 500mm high.	1 Horizontal. 2 Vertical. 3 Raking. 4 Curved, radius stated. 5 Stepped. 6 Preformed. 7 Sloping.	1 Boundary work to voids is only measured where the void is >1.00m². 2 Boundary work is work associated with closing off or finishing off sheet roofing at the external perimeter, at the abutment with different materials or the perimeter of openings and voids. 3 Boundary work is deemed to include undercloaks, insulation, strip ventilators, rough and fair cutting, bedding, pointing, laps, seams, ends, angles, intersections, rolls, upstands, downstands, welted edges, dressings and wedgings and additional covering material needed to form the detail and all associated labours. 4 Where several items of the same type of boundary work have slightly differing net girths these girths may be averaged and the lengths aggregated. 5 The number of separate lengths aggregated must be stated 6 Valleys include everything necessary to form and line the valley. 7 An upstand or downstand over 500mm high is measured as vertical work.

Item or work to be measured	Unit	Level one	Level two	Level three	Notes, comments and glossary
5 Flashings	m	1 Net girth stated.	1 Flashings. 2 Aprons. 3 Sills. 4 Weatherings. 5 Cappings. 6 Hips. 7 Kerbs. 8 Ridges. 9 Linings to openings.	1 Horizontal. 2 Sloping 3 Vertical. 4 Raking. 5 Curved, radius stated. 6 Stepped. 7 Preformed.	1 Flashings are deemed to include undercloaks, rough and fair cutting, bedding, pointing, ends, angles, intersections, welted, beaded or shaped edges and all dressings.
6 Gutters 7 Valleys	m	1 Net girth stated.	1 Sloping. 2 Stepped. 3 Curved, radius stated. 4 Secret. 5 Tapered.	1 Nature of base. 2 Spacing of structural supports. 3 Preformed.	1 The length is the mean length measured over all fittings. 2 Gutter and valley work is deemed to include all dressings required to form the profile complete with all joints, laps, seams, brackets, undercloaks and other associated linings, outlets, overflows, ends, angles, intersections, bedding, pointing, fixings and the like. 3 Maximum and minimum girth of tapered gutters or valleys shall be given.

Item or work to be measured	Unit	Level one	Level two	Level three	Notes, comments and glossary
8 Spot items	nr	1 Dimensioned diagram or dimensioned description.	1 Catchpits. 2 Sumps. 3 Outlets. 4 Hatch covers. 5 Canopy covers. 6 Collars or sleeves around pipes and the like. 7 Other: type stated.	1 Nature of base. 2 Spacing of structural supports. 3 Preformed.	1 Spot item work is deemed to include joints, dressing and bonding to surrounding work, undercloaks and other associated linings, ends, angles, bedding, pointing, fixings and the like.
9 Fittings	nr	1 Dimensioned description.	1 Ventilators. 2 Finials. 3 Gas terminals. 4 Hip irons. 5 Soakers. 6 Saddles. 7 Rooflights. 8 Other: type stated.	1 Nature of base. 2 Method of fixing where not at the discretion of the contractor.	1 Fittings are deemed to include joints, dressing and bonding to surrounding work, undercloaks and other associated linings, ends, angles, bedding, pointing, fixings and the like. 2 Proprietary references may be given in lieu of a dimensioned description.

18 Tile and slate roof and wall coverings

Plain tiling
Interlocking tiling
Fibre cement slating
Natural slating
Natural or artificial stone slating
Timber or bituminous felt shingles
Any other type of tile, slate, slab or block roof or wall covering

Drawings that must accompany this section of measurement.	Mandatory information to be provided.	Notes, comments and glossary
1 Plans of each roof. 2 Principal sections of roof. 3 External elevations.	1 Kind, quality and size of materials. 2 Method of fixing. 3 Minimum laps. 4 Spacing of battens and counter battens. 5 Composition and mix of mortar. 6 Type of pointing. 7 Bonding to other work. 8 Radius of curved work.	
Minimum information that must be shown on the drawings that accompany this section of measurement.	**Works and materials deemed included.**	
1 Major horizontal and vertical dimensions. 2 Types of materials used to construct the roof and its structures. 3 Position of main structural frame members. 4 Height of work above ground level.	1 All rough and fair square, raking and curved cutting. 2 All ends and angles either formed or proprietary. 3 Extra material for curved work. 4 Forming all rough and fair grooves, throats, mortices, chases, rebates, holes, stops, mitres and the like labours. 5 Raking out joints to form a key.	

Item or work to be measured	Unit	Level one	Level two	Level three	Notes, comments and glossary
1 Roof coverings 2 Wall coverings	m²	1 Pitch stated. 2 Vertical.	1 Underlays and battens.	6 Raking out joints to insert flashings. 7 Labours in returns, ends and angles. 8 All extra material required for bonding. 9 Additional material in laps. 10 Work in forming voids and holes ≤ 1.00m². 1 Curved: radii stated. 2 Conical: maximum and minimum radii stated.	1 Coverings are deemed to include underlays, battens and work in forming voids ≤ 1.00m². 2 No deduction is made for voids ≤ 1.00m².
3 Boundary work; location and method of forming described	m	1 Dimensioned description stating net girth.	1 Abutments. 2 Eaves. 3 Ridges. 4 Verges. 5 Valleys. 6 Hips. 7 Vertical angles.	1 Horizontal. 2 Sloping. 3 Raking. 4 Vertical. 5 Curved: radius stated. 6 Stepped. 7 Pre formed.	1 Boundary work is deemed to include undercloaks, rough and fair cutting, bedding, pointing, ends, angles and intersections. 2 Boundary work to voids is only measured where the void exceeds 1.00m². 3 Boundary work is work associated with closing off or finishing off tile or slate roofing at the external perimeter; at the abutment with different materials or the perimeter of openings and voids. 4 Valleys include everything necessary to form and line the valley excluding any sheet metal. Sheet metal linings shall be measured in Work Section 17: Sheet Roof coverings.

Item or work to be measured	Unit	Level one	Level two	Level three	Notes, comments and glossary
4 Fittings	nr	1 Dimensioned description.	1 Ventilators. 2 Finials. 3 Gas terminals. 4 Hip irons. 5 Soakers. 6 Saddles. 7 Rooflights. 8 Other: type stated.	1 Nature of base. 2 Method of fixing where not at the discretion of the contractor.	1 Proprietary references may be given in lieu of a dimensioned description.

19 Waterproofing

Mastic asphalt roofing
Applied liquid roofing
Asphalt tanking or damp proofing
Applied liquid tanking or damp proofing
Flexible sheet tanking or damp proofing
Other proprietary systems of tanking or damp proofing

	Mandatory information to be provided.	Notes, comments and glossary
Drawings that must accompany this section of measurement.	1 Plans of each level where work is to be carried out. 2 Sections showing extent of work.	1 Restrictions on siting of plant and materials. 2 Kind and quality of all materials. 3 Thickness and number of coats. 4 Surface finishes or treatments.
Minimum information that must be shown on the drawings that accompany this section of measurement.	1 Extent of work. 2 Height of work above ground level. Works and materials deemed included.	1 Cutting to line. 2 All labour and material associated with cutting, notching, bending, lapping and reinforcement. 3 Working into recesses such as duct covers, shaped insets, manhole covers, mat wells, outlet pipes, dished gullies and the like. 4 Work to falls and cross falls. 5 All boundary work to openings ≤ 1.00m² 6 All preparation to background necessary to form a key, including raking out joints, scabbing or the application of a bonding agent.

Item or work to be measured	Unit	Level one	Level two	Level three	Notes, comments and glossary
1 Coverings > 500mm wide	m²	1 Horizontal. 2 Sloping: pitch stated. 3 Vertical. 4 Curved: radii stated.	1 Underlays. 2 Insulation. 3 Finish to exposed surface. 4 Protection.	1 Nature of base. 2 Number of coats or layers.	1 The area measured is that in contact with base. 2 No deduction is made for voids ≤ 1.00m².
2 Coverings ≤ 500mm wide 3 Skirtings 4 Fascias 5 Aprons	m	1 Net girth on face.	1 Horizontal. 2 Sloping. 3 Vertical. 4 Raking. 5 Curved: radius stated. 6 Stepped.		1 Boundary work to voids is only measured where the void is > 1.00m². 2 Boundary work is work associated with closing off or finishing off the waterproofing at the external perimeter, at the abutment with different materials or the perimeter of openings and voids. 3 Boundary work is deemed to include undercloaks, insulation, strip ventilators, rough and fair cutting, bedding, pointing, laps, seams, ends, angles, intersections, rolls, upstands, downstands, dressings and wedgings.
6 Gutters 7 Channels 8 Valleys 9 Kerbs	m	1 Net girth on face.	1 Sloping. 2 Stepped. 3 Curved. 4 Secret. 5 Tapered.	1 Nature of base. 2 Number of coats or layers. 3 Spacing of structural supports.	1 The length is the mean length measured over all fittings. 2 Lining and covering work is deemed to include all dressings required to form the profile, joints, brackets, undercloaks and other associated linings, outlets, overflows, ends, angles, intersections, bedding, pointing, fixings and the like. 3 Maximum and minimum girth of tapered gutters shall be given.

Item or work to be measured	Unit	Level one	Level two	Level three	Notes, comments and glossary
10 Spot items	nr	1 Dimensioned diagram or dimensioned description.	1 Catch pits. 2 Sumps. 3 Outlets. 4 Hatch covers. 5 Canopy covers. 6 Collars or sleeves around pipes and the like. 7 Other: type stated.		1 Spot item work is deemed to include joints, dressing and bonding to surrounding work, undercloaks and other associated linings, ends, angles bedding, pointing, fixings and the like.
11 Fittings	nr	1 Dimensioned description.	1 Ventilators. 2 Other: type stated.	1 Nature of base. 2 Number of coats or layers. 3 Method of fixing where not at the discretion of the contractor.	1 Fittings are deemed to include joints, dressing and bonding to surrounding work, undercloaks and other associated linings, ends, angles bedding, pointing, fixings and the like. 2 Proprietary references may be given in lieu of a dimensioned description.
12 Edge trim	m				

20 Proprietary linings and partitions

Metal framed systems to walls and ceilings
Drylining and partitioning systems to walls and ceilings

		Mandatory information to be provided.	Notes, comments and glossary
Drawings that must accompany this section of measurement.	Plans, sections and elevations to show scope and location of work.	1 Kind, quality and size or thickness of materials and components. 2 Method of fixing framing and linings. 3 Layout and treatment of joints. 4 Method of jointing. 5 Surface preparation. 6 Finish. 7 Radii of curved work. 8 Nature of base. 9 Location of wall or ceiling mounted fittings requiring additional support or framing. 10 Sealants to joints and perimeters.	1 The finish is that which is applied as part of the normal proprietary system. Any finish to be subsequently made by a different trade shall be measured separately under the relevant rules. 2 Examples of proprietary finishes are taped joints, slurry coats, and the like. 3 Skim coats or other wet trade finishes are measured in Work Section 28: Floor, wall, ceiling and roof coverings.
Minimum information that must be shown on the drawings that accompany this section of measurement.	1 Structural floor to ceiling heights. 2 The services located within the ceiling, partition or lining where the work includes complex integral services.	Works and materials deemed included. 1 All work is deemed internal unless otherwise stated. 2 All rough and fair cutting. 3 Extra material for curved work. 4 Additional framing to support fittings. 5 Ends, fair ends and abutments with adjoining work. 6 Working around columns, beams and services where the outer face of the work is continuous.	1 This refers to the final location of the work.

Item or work to be measured	Unit	Level one	Level two	Level three	Notes, comments and glossary
1 Proprietary metal framed system to form walls	m²	1 Finished thickness stated 2 Height or average height stated in 1.00m increments. 3 Total length stated measured along centre line.	1 Insulation. 2 Vapour barriers. 3 Sub linings. 4 Finish. 5 Glazing 6 Curved: radius stated.		1 No deductions for voids ≤ 1.00m². 2 The average height will be calculated for each length of partition with a sloping head measured between junctions.
2 Proprietary metal framed system to form ceilings	m²	1 Over 300mm wide on face	1 Insulation. 2 Vapour barriers. 3 Sub linings. 4 Finish. 5 Curved: radius stated. 6 Sloping 7 Convex or concave: radius stated.	1 Fixed direct to structural soffit 2 Supported on adjacent structure, span stated in 1.00m increments	1 The adjacent structure such as secondary steel framing or timber framing is measured elsewhere.
	m	1 Not exceeding 300mm wide on face			
3 Extra over for different	m²	1 Lining: details stated. 2 Finish: details stated.			1 This will apply to partitions that have areas of different linings or finishes than that specified in the general description heading. It will not apply to different forms of construction or components. These differences will require separate items measured.

Item or work to be measured	Unit	Level one	Level two	Level three	Notes, comments and glossary
4 Extra over for forming openings	nr	1 Not exceeding 2.50m². 2 2.50–5.00m². 3 Exceeding 5.00m² in further increments of 2.50m².	1 Lined: details stated. 2 Unlined.		1 Non-proprietary materials are not deemed included (e.g. timber grounds or inserts would be measured in Work Section 16: Carpentry). 2 Lined openings are those lined as part of this work.
5 Extra over for non-standard perimeter details	m	1 Dimensioned description or proprietary reference.	1 Heads: details stated. 2 Soles: details stated. 3 Abutments: details stated.	1 Nature of base.	1 Non-standard details are those that involve the use of components that are not used within the main body of work such as deflection heads, acoustic seals, fire seals and the like.
6 Extra over for angles	m				
7 Extra over for junctions	m		1 Tee. 2 Cross.		
8 Extra over for access panels	nr	1 Dimensioned description or proprietary reference.	1 Details stated.		1 Details include type of frame, ironmongery, finish, insulation, fire rating, and proprietary reference number where applicable. 2 Fair ends are only measured where the exposed end is finished with the same finish as the face(s) or with a trim which is an integral part of the partition system. 3 Fair ends are deemed to include all extra work and materials involved.
9 Fair ends to partitions	m	1 Thickness of partition stated.	1 Finish and/or trims.		

Item or work to be measured	Unit	Level one	Level two	Level three	Notes, comments and glossary
10 Proprietary linings to walls	m²	1 Over 300mm wide on face. 2 Not exceeding 300mm wide on face.	1 Insulation. 2 Vapour barriers. 3 Sub linings. 4 Finish. 5 Curved, radius stated. 6 Sloping. 7 Convex or concave: radius stated.	1 Method of fixing 2 Nature of base	
11 Proprietary linings to ceilings	m²				
12 Proprietary linings to columns	m	1 Girth n/e 300mm, nr of faces stated. 2 Girth 300–600mm, nr of faces stated. 3 Girth 600–900mm, nr of faces stated. 4 Thereafter in 300mm stages, nr of faces stated.			
13 Proprietary linings to beams	m				
14 Proprietary linings to bulkheads	m				
15 Extra over for forming openings	nr	1 Not exceeding 2.50m². 2 2.50–5.00m². 3 Exceeding 5.00m² in further increments of 2.50m².	1 Lined, details stated. 2 Unlined.		1 This includes openings for doors, windows, screens and the like.
16 Extra over for non-standard perimeter details	m	1 Dimensioned description or proprietary reference.	1 Heads: details stated. 2 Soles: details stated. 3 Abutments: details stated.	1 Nature of base.	1 Non-standard details are those that involve the use of components that are not used within the main body of work such as deflection heads, acoustic seals, fire seals and the like.
17 Extra over for angles.	m				
18 Extra over for junctions.	m		1 Tee. 2 Cross.		

Item or work to be measured	Unit	Level one	Level two	Level three	Notes, comments and glossary
19 Extra over for access panels.	nr	1 Dimensioned description or proprietary reference.	1 Details stated.		1 Details include type of frame, ironmongery, finish, insulation, fire rating and proprietary reference number where applicable.
20 Beads, function stated	m		1 Method of fixing.	1 Nature of base.	1 Function of beads include angle beads, stop beads, shadow gap beads, casing beads and the like.

21 Cladding and covering

Patent glazing, curtain walling, rainscreen cladding, glazed vaulting and structural glass assemblies
Rigid sheet cladding
Weatherboarding
Profiled sheet cladding or roofing
Panel or slab cladding or roofing
Sheet claddings or coverings
Any other type of cladding, lining or covering

		Mandatory information to be provided.	Notes, comments and glossary
Drawings that must accompany this section of measurement.	1 Plans, sections and elevations sufficient to show scope and location of the works. 2 Component drawings.	1 Kind and quality of materials. 2 Type, finish and spacing of framing members. 3 Nature, thickness and spacing of structural supports.	
Minimum information that must be shown on the drawings that accompany this section of measurement.	1 Construction of each installation.	Works and materials deemed included. 1 Ironmongery where supplied with the component. 2 Glazing where part of the installation. 3 Integral insulation, membranes and the like where part of the system. 4 Mastics and sealants. 5 Cleats and brackets. 6 Fixings and fastenings. 7 Mechanical and electrical operating equipment where supplied with the component.	1 Secondary steel support work is measured in accordance with the rules of Work Section 15: Structural Metalwork.

Item or work to be measured	Unit	Level one	Level two	Level three	Notes, comments and glossary
1 Walls 2 Floors 3 Ceilings 4 Roofs 5 Sides and tops of dormers 6 Sides and soffits of beams 7 Sides of columns	m	1 Not exceeding 600 wide.	1 Laid diagonally. 2 Sloping; pitch stated. 3 Vertical. 4 Curved: radius stated. 5 To soffits.	1 Internal. 2 External.	
1 Walls 2 Floors 3 Ceilings 4 Roofs 5 Sides and tops of dormers 6 Sides and soffits of beams 7 Sides of columns	m²	1 Exceeding 600 wide.			
	nr	1 Isolated area not exceeding 1 m², irrespective of width.			
8 Items extra over the work in which they occur	nr	1 Dimensioned description. 2 Proprietary reference.	1 Opening lights. 2 Doors. 3 Rooflights. 4 Ventilator panels. 5 Forming openings for other trades.	1 Electrical requirements.	1 These items are only for components that are part of the system to which they are installed.
9 Boundary work	m		1 Heads. 2 Ridges. 3 Valleys. 4 Hips. 5 Bottom edges. 6 Eaves. 7 Verges. 8 Abutments. 9 Flashing pieces.	1 Horizontal. 2 Vertical. 3 Sloping. 4 Raking. 5 Curved, radius stated. 6 Stepped. 7 Preformed.	1 All edge and intermediate trims, cover pieces, mastics and sealants are deemed included with the associated boundary or opening perimeter work. 2 Boundary work is work associated with closing off or finishing off claddings or coverings at the external perimeter or at the abutment with different materials.

Item or work to be measured	Unit	Level one	Level two	Level three	Notes, comments and glossary
10 Opening perimeters	m	1 Dimensioned description. 2 Proprietary reference.	1 Heads. 2 Jambs. 3 Sills. 4 Flashing pieces.	1 Horizontal. 2 Vertical. 3 Sloping. 4 Raking. 5 Curved, radius stated. 6 Stepped. 7 Preformed.	1 opening work is associated with closing off or finishing off claddings and coverings at the perimeter of an opening or a void.
11 Angles	m		1 Internal. 2 External.	1 Regular. 2 Irregular: angle stated.	1 Irregular angles are any that are not 90°.
12 Closers	m		1 Fire stops. 2 Other: details stated.		

22 General joinery

Unframed isolated trims, skirtings, or sundry joinery items

In-fill panels and sheets

Sealant joints

General ironmongery not associated with windows and doors

Drawings that must accompany this section of measurement.		Mandatory information to be provided.	
	1 General arrangement plans.		1 Kind and quality of materials.
			2 Finish of timber if not wrot.
			3 Preservation treatments.
			4 Surface treatments as part of the production process.
			5 Selection and protection for subsequent treatment or clear finish.
			6 Matching grain or colour.
			7 Limits on planing margins.
			8 Method of jointing and construction where not at the discretion of the contractor.
			9 Form of construction and jointing where not at the discretion of the contractor.
			10 Method of fixing where not at the discretion of the contractor.
Minimum information that must be shown on the drawings that accompany this section of measurement.		Works and materials deemed included.	
	1 Scope and location of work.		1 All timbers are deemed to be wrot finish unless described otherwise.
			2 All timber sizes are nominal unless stated as finished size.
			3 Ends, angles, mitres and intersections irrespective of cross section area of timber.
			4 Work is deemed internal unless described as external.

Notes, comments and glossary

Item or work to be measured	Unit	Level one	Level two	Level three	Notes, comments and glossary
Unframed isolated trims, skirtings and sundry items					
1 Skirtings, picture rails 2 Architraves and the like 3 Cover fillets, stops, trims, beads, nosings and the like 4 Isolated shelves and worktops 5 Window boards 6 Isolated handrails and grab rails	m	1 Dimensioned overall cross section stated. 2 Number and type of labours described.	1 Built up members, size of components stated. 2 Timber components tongued on. 3 Different cross-section shapes (nr).	1 Fixing through vulnerable materials. 2 Curved work to be described stating the radius.	
7 Duct covers 8 Pipe casings 9 Shelves	m	1 Thickness and width or girth stated. 2 Thickness and width or girth not exceeding 300mm stated. 3 Thickness and width or girth over 300mm but not exceeding 600mm stated. 4 Thickness and width or girth thereafter in 300mm stages.	1 Nature and method of forming joints.		1 Usually the exact width or girth shall be given for these items but where there are many differing but similar widths these may be grouped into 300mm wide or girth bands.
10 Pinboards, backboards, plinth blocks and the like	nr	1 Dimensioned description.		1 Method of fixing. 2 Nature of base.	
Floor, wall and ceiling boarding, sheeting, panelling, linings and casings					
11 Boarding, sheeting, panelling over 600mm wide	m²	1 Thickness.	1 Floors. 2 Walls. 3 Ceilings.	1 Method of fixing. 2 Method of jointing. 3 Nature of base.	
12 Boarding, sheeting, panelling not exceeding 600mm wide	m	1 Width and thickness.			

Item or work to be measured	Unit	Level one	Level two	Level three	Notes, comments and glossary
Proprietary partitions, panels and cubicles					
13 Partitions	m	1 Height and thickness stated. 2 Proprietary name or reference stated.	1 Off site applied finish. 2 On site applied finish.	1 Method of jointing and fixing. 2 Curved: radius stated. 3 Nature of background. 4 Integral services.	1 The linear measurement is the mean length measured along the centre line of the partition. 2 The length is measured through and over all obstructions and openings. 3 Partitions are deemed to include all integral components, fixings, joints, factory applied trims, holes and mortices.
14 Items extra over the partition they occur in	nr	1 Openings: size stated.	1 Blanks. 2 Doors. 3 Windows. 4 Glazed panels. 5 Access panels.		1 Openings are deemed to include everything necessary to form the opening together with their associated integral components, glass, doors, ironmongery, linings and factory-applied trims.
	m	2 Trims: dimensioned description.			1 Trims are only measured as separate items if fixed on site.
15 Duct panels, Sanitary ware back panels and the like	nr	1 Overall size stated. 2 Thickness stated.	Sub frames, details stated.	1 Method of fixing 2 Nature of background.	1 All ancillary items needed to connect the sanitary appliances to the M and E installations are deemed included. 2 Associated sanitary ware items are not included in these items but shall be measured in accordance with the rules of Work Section 32: Furniture, fittings and equipment
16 Cubicle partition sets	nr	1 Dimensioned diagram or dimensioned description.	1 Complete cubicles: number of bays stated.	1 Nature of background. 2 Method of fixing to background.	1 Cubicle sets are deemed to include all integral frames, panels, doors, factory applied trims, connections, fixings and fastenings complete with support legs, brackets and standard ironmongery.

Item or work to be measured	Unit	Level one	Level two	Level three	Notes, comments and glossary
17 Items extra over the partition they occur in	m	1 Trims: dimensioned description.			1 Trims are only measured as separate items if fixed on site.
Infill panels					1 Infill panels are non-glass and non-glass plastics rigid sheet spandrel and infill panels of all kinds fixed with beads, gaskets and the like to wood, metal, plastic and concrete surrounds. They exclude panels or sheets forming an integral part of a component or proprietary cladding system.
18 Infill panels and sheets, number stated	m²	1 Thickness stated.	1 Curved: stating radius. 2 Panels exceeding size of normal manufactured unit. 3 Panels requiring special treatment to edges.		1 Work is deemed internal unless described as external. 2 Infill panels are deemed to include bedding compounds, sealants, intumescent compounds and strips, distance pieces, location and setting blocks and fixings.
Sealant joints					1 Kind and quality of materials. 2 Method of application. 3 Method of preparing contact surfaces.
19 Joints: contact surfaces stated	m	1 Type and size of components stated.	1 Vertical. 2 Sloping. 3 Soffit. 4 Horizontal.		1 Lengths are measured on face. 2 Work is deemed to include preparation, cleaners, primers, sealers, backing strips and fillers.
20 Pointing: contact surfaces stated	m	1 One side. 2 Both sides.			

Item or work to be measured	Unit	Level one	Level two	Level three	Notes, comments and glossary
21 Raking out existing joints	m	1 Width , depth and type of material stated.	1 Vertical. 2 Sloping. 3 Soffit. 4 Horizontal.		3 Deemed to include disposal of all debris.
Ironmongery					1 Kind and quality of materials and fixings. 2 Surface finish. 3 Constituent parts of the units or sets. 4 Fixing through vulnerable materials.
22 Type of item, unit or set stated	nr	1 Method of fixing. 2 Nature of base.			1 Ironmongery is deemed to include fixing with screws to match and preparing the base to receive it. 2 Shelf brackets and associated fittings will be measured here.

23 Windows, screens and lights

Timber, metal, plastic:
Windows and shop fronts
Rooflights
Screens
Louvres, shutters, canopies and blinds
Associated glass and glazing
Associated ironmongery

Drawings that must accompany this section of measurement.	Mandatory information to be provided.	Notes, comments and glossary
1 General arrangement plans. 2 Door and window schedule. 3 Glazing schedule. 4 Ironmongery schedule.	1 Kind and quality of materials and if timber describing if wrot or sawn 2 Preservation treatments. 3 Surface treatments applied as part of the production process. 4 Selection and protection for subsequent treatment or clear finish. 5 Matching grain or colour. 6 Limits on planing margins. 7 Method of jointing and construction where not at the discretion of the contractor. 8 Form of construction and jointing where not at the discretion of the contractor. 9 Method of fixing where not at the discretion of the contractor. 10 Background where vulnerable.	
Minimum information that must be shown on the drawings that accompany this section of measurement.	**Works and materials deemed included.**	
1 Shape and size of units. 2 Methods of fixing units. 3 Glazing requirements.	1 Bedding and pointing frames. 2 Glass and the like is deemed to be installed on site unless stated otherwise.	

Item or work to be measured	Unit	Level one	Level two	Level three	Notes, comments and glossary
1 Windows and window frames	nr	1 Dimensioned description or diagram.	1 Curved work: radius stated.	1 Method of fixing if not left to discretion of contractor. 2 Factory glazed.	1 Glass may be incorporated into the associated work following the rules set out below.
2 Window shutters					
3 Sun shields					
4 Rooflights, skylights and lanternlights					
5 Screens, borrowed lights and frames					
6 Shop fronts					
7 Louvres and frames					
Glazing					
8 Glass, type stated	nr	1 Thickness of glass or overall thickness of sealed unit stated. 2 Pane size.	1 Shape if other than square or rectangular. 2 Airspace(s) width(s) of double/triple glazed units. 3 Gas filling of airspaces.	1 Bent: direction stated. 2 Method of glazing or securing. 3 Edge treatment.	1 Glass includes plastic, or any type of material glazed into openings except glass blocks. 2 The size given for irregular panes shall be the smallest rectangular size from which the pane can be obtained.
9 Louvre blades					
Ironmongery					
10 Type of item, unit or set stated	nr	1 Method of fixing. 2 Nature of base.	1 Kind and quality of materials and fixings. 2 Surface finish. 3 Constituent parts of the units or sets.		1 Ironmongery is deemed to include fixing with screws to match and preparing the base to receive it.

24 Doors, shutters and hatches

Timber, metal, plastic:
Doors and frames
Shutters and Hatches
Sliding or folding doors and partitions
Grilles
Associated glass and glazing
Associated ironmongery

		Notes, comments and glossary
Drawings that must accompany this section of measurement.	1 Floor plans. 2 Door schedule. 3 Ironmongery schedule.	Mandatory information to be provided.
		1 Kind and quality of materials and, if timber, describing if wrot or sawn. 2 Preservation treatments. 3 Surface treatments applied as part of the production process. 4 Selection and protection for subsequent treatment or clear finish. 5 Matching grain or colour. 6 Limits on planing margins. 7 Method of jointing and construction where not at the discretion of the contractor. 8 Form of construction and jointing where not at the discretion of the contractor. 9 Method of fixing where not at the discretion of the contractor. 10 Background where vulnerable.
Minimum information that must be shown on the drawings that accompany this section of measurement.	1 Location of all doors. 2 Door reference number.	Works and materials deemed included.
		1 Bedding and pointing frames. 2 Glass and the like is deemed to be installed on site unless stated otherwise.

Item or work to be measured	Unit	Level one	Level two	Level three	Notes, comments and glossary
1 Door sets	nr	1 Detailed description of set. 2 Structural opening size.	1 Background.	1 Smoke stops: details stated. 2 Fire stops: details stated.	1 Door sets comprise the door or doors complete with associated frame, stops, architraves and trims. 2 Any trims and the like fixed to the set after installation must be measured in accordance with the rules of Work Section 22: General joinery. 3 Glass may be incorporated into the associated work following the rules set out below.
2 Doors	nr	1 Dimensioned description or diagram.	1 Fire resistance performance.	1 Method of fixing if not left to discretion of contractor 2 Factory glazed	1 Each leaf of a multi-leafed door is counted as one door. 2 Glass may be incorporated into the associated work following the rules set out below.
3 Roller shutters					
4 Collapsible gates					
5 Sliding folding partitions					
6 Hatches					
7 Strong room doors					
8 Grilles					
9 Door frames	m	1 Dimensioned overall cross section description.	1 Labours described.		1 Frames and linings includes jambs, heads, sills, mullions and transomes.
10 Door linings					
11 Door stops					
12 Associated fire stops					
13 Associated smoke stops					

Item or work to be measured	Unit	Level one	Level two	Level three	Notes, comments and glossary
Glazing					
14 Glass, type stated	nr	1 Thickness of glass or overall thickness of sealed unit stated. 2 Pane size.	1 Shape if other than square or rectangular. 2 Airspace(s) width(s) of double/triple glazed units. 3 Gas filling of airspaces.	1 Bent: direction stated. 2 Method of glazing or securing. 3 Edge treatment.	1 Glass includes plastic, or any type of material glazed into openings except glass blocks. 2 The size given for irregular panes shall be the smallest rectangular size from which the pane can be obtained.
15 Louvre blades					
Ironmongery					
16 Type of item, unit or set stated	nr	1 Method of fixing. 2 Nature of base.	1 Kind and quality of materials and fixings. 2 Surface finish. 3 Constituent parts of the units or sets.		1 Ironmongery is deemed to include fixing with screws to match and preparing the base to receive it.

25 Stairs, walkways and balustrades

Timber, metal, plastic:
Staircases
Walkways and Gantries
Balustrades
Barriers
Guardrails

Drawings that must accompany this section of measurement.		Mandatory information to be provided.	Notes, comments and glossary
1 Plans and sections showing location and scope of each unit.		1 Kind and quality of materials and if timber describing if wrot or sawn.	
		2 Preservation treatments.	
		3 Surface treatments applied as part of the production process.	
		4 Selection and protection for subsequent treatment or clear finish.	
		5 Matching grain or colour.	
		6 Limits on planing margins.	
		7 Method of jointing and construction where not at the discretion of the contractor.	
		8 Form of construction and jointing where not at the discretion of the contractor.	
		9 Method of fixing where not at the discretion of the contractor.	
		10 Background where vulnerable.	

Minimum information that must be shown on the drawings that accompany this section of measurement.		Works and materials deemed included.		Notes, comments and glossary
1 Principle dimensions of each staircase or unit. 2 Height from structural floor to structural floor level.		1 All ramps, wreaths, bends, plain ends, ornamental ends and the like labours. 2 Linings, nosings, cover moulds, trims and the like where integral part of the unit. 3 Soffit linings, spandrel panels and the like where an integral part of the unit. 4 Ironmongery where an integral part of the unit. 5 Finishes applied off site. 6 Fixings, fastenings, blockings, wedges, bolts, brackets, cleats and the like.		1 Staircases are deemed to include attached balustrades and newel posts. 2 Any trims and the like fixed to the unit after installation must be measured in accordance with the rules of Work Section 22: General joinery.

Item or work to be measured	Unit	Level one	Level two	Level three	Notes, comments and glossary
1 Staircase: type stated	nr	1 Dimensioned description. 2 Component drawing reference.		1 Hatch doors where part of the loft ladder component.	
2 Loft ladders					
3 Ladders					
4 Extra over for:	nr	1 Quarter landing. 2 Half landing.	1 Dimensioned description.		
5 Catwalks	m	1 Dimensioned description.	1 Curved; radius stated.	1 Method of fixing or support. 2 Background.	
6 Walkways					

Item or work to be measured	Unit	Level one	Level two	Level three	Notes, comments and glossary
7 Balustrades	m - - - - nr	1 Dimensioned description. - - - - - - - - - - - - - - - - 2 Component drawing reference.	1 Curved: radius stated.	1 Infill panels: detail stated. 2 Method of fixing or support. 3 Background.	1 These items are measured when not forming an integral part of a staircase unit. 2 These items include all integral rails, infill panels, ironmongery, factory applied finishes, fixings and fastenings.
8 Handrails					
9 Barriers					
10 Guard rails					
11 Balcony units					
12 Extra over for:	nr	1 Opening portions: size and details stated.			

26 Metalwork

Isolated metal members
General metalwork
General metalwork fittings and fixtures

			Mandatory information to be provided.	Notes, comments and glossary
Drawings that must accompany this section of measurement.		1 Plans, sections, elevations or details sufficient to show location, scope and construction of each item measured.	1 Types and grades of materials. 2 Details of connections. 3 Details of welding tests and x-rays. 4 Details of performance tests.	
Minimum information that must be shown on the drawings that accompany this section of measurement.		1 Position of the work. 2 Types and sizes of structural members and their position relative to each other.	Works and materials deemed included. 1 All fabrication and erection. 2 All bolts, nuts, washers, fixings and fastenings.	

Item or work to be measured	Unit	Level one	Level two	Level three	Notes, comments and glossary
1 Isolated metal members	m	1 Dimensioned description: cross section profile stated.	1 Built up: details of construction stated. 2 Tapered. 3 Curved: radius stated. 4 Hollow: shape stated.	1 Method of fixing in position. 2 Fixing plates, brackets and the like: dimensioned description stated. 3 Protective coating applied off site.	1 Forming holes, mortices and the like in the item and background for fixing purposes are deemed included.
2 General metalwork members	m nr	1 Dimensioned description: cross section profile stated.	1 Protective coating applied off site.	1 Method of fixing. 2 Nature of background. 3 Surface pattern or finish.	
3 Sheet metal	m² nr	1 Thickness stated. 2 Dimensioned description.			

Item or work to be measured	Unit	Level one	Level two	Level three	Notes, comments and glossary
4 Wire mesh	m² - - - - nr	1 Mesh size and thickness stated. - - - - - - - - - - - - 2 Dimensioned description.	1 Protective coating applied off site.	1 Method of fixing. 2 Nature of background. 3 Surface pattern or finish.	1 Forming holes, mortices and the like in the item and background for fixing purposes are deemed included.
5 Composite items	m - - - - nr	1 Dimensioned description. 2 Drawing reference.	1 Full description of all components including integral non-metallic items.	1 Method of fixing. 2 Nature of background.	1 Non-metallic integral items forming part of the composite item may be included in the composite item provided they are clearly described as such. 2 Any non-metallic item applied after the composite item has been installed on site shall be measured in accordance with the rules of the relevant section.
6 Filling hollow sections	item	1 Water. 2 Concrete. 3 Other material: type stated.	1 Details stated.		
7 Surface treatments	m²	1 Galvanising. 2 Sprayed coating. 3 Painting. 4 Other treatment: type stated.	1 On site. 2 Off site.	1 Surface preparation described.	1 All preparation is deemed included. 2 Any finish applied after installation is measured in accordance with the rules of the 'Painting and decorating' section.
8 Isolated protective coatings	nr	1 Approximate size or area stated	1 Type of protective coated stated. 2 Preparation described.	1 On site. 2 Off site.	

27 Glazing

Drawings that must accompany this section of measurement.				Mandatory information to be provided.	Notes, comments and glossary
1 Elevations. 2 Component drawings.					1 This section covers all types of glass and glass plastics and glazing with putty, beads or gaskets into prepared openings. 2 Glazing supplied as part of a window, door or other component is measured elsewhere with those components.

Minimum information that must be shown on the drawings that accompany this section of measurement.					
1 General scope and location of work.				1 Works and materials deemed included.	

Item or work to be measured	Unit	Level one	Level two	Level three	Notes, comments and glossary
1 Glass, type stated 2 Sealed glazed units, type of glass stated 3 Louvre blades, type of glass stated	nr	1 Thickness of glass or overall thickness of sealed unit. 2 Pane size.	1 Shape if other than square or rectangular. 2 Airspace(s) width(s) of double/triple glazed units. 3 Gas filling of airspaces.	1 Kind quality and thickness of glass. 2 Kind and quality of glazing material. 3 Method of glazing including details of gaskets where required. 4 Nature of frame or background.	1 Glass includes laminated, toughened, double or triple glazing units, lead, acrylic, polycarbonate and the like, or any other type of material glazed into openings except glass blocks. 2 Glass blocks are measured in accordance with the rules of Work Section 14: Masonry. 3 The size given for irregular panes shall be the smallest rectangular size from which the pane can be obtained.
				1 All raking and curved cutting. 2 All polished and bevelled edges unless given in the description. 3 Drilling holes.	
				Level three	
			Level two	1 Bent: direction stated. 2 Method of glazing or securing: details stated. 3 Edge treatment.	
4 Extra for:	nr	1 Grinding. 2 Sandblasting. 3 Acid etching. 4 Embossing. 5 Engraving.	1 Plain work; pane size stated. 2 Design work; pane size stated.	1 Detailed description of work including extent of work over the pane.	1 This information may be given by reference to detailed drawings.

EFFECTIVE FROM 1 JANUARY 2013

Item or work to be measured	Unit	Level one	Level two	Level three	Notes, comments and glossary
5 Lead light glazing, type of glass stated	nr	1 Thickness of glass. 2 Overall window or light size.	1 Pane sizes where regular shape. 2 Pattern of lead cames. 3 Spacings of lead cames.	1 Bent: direction stated. 2 Method of glazing or securing: details stated. 3 Edge treatment.	1 The pattern will be described as regular, diamond or irregular. 2 Cames are the H-shaped lead strips that are soldered together to form the lead glazed panes.
6 Saddle bars	m	1 Diameter and length stated.	1 Method of fixing to window surround.	1 Method of attaching to lead light glazing.	1 Saddle bars are metal rods used to strengthen and provide support to large lead glazed panes. These bars are usually built in to the window surround.
7 Mirrors	nr	1 Dimensioned description.	1 Edge description. 2 Pattern: method of forming described.	1 Nature of background. 2 Method of fixing: details stated.	1 Edges can be plain, rounded, bevelled and the like.
8 Removing existing glass and preparing frame or surround to receive new glass	nr	1 Type of frame or surround. 2 Type of glass to be removed.	1 Panes $\leq$ 1.00m^2. 2 Panes 1.00–2.00m^2. 3 Panes 2.00–3.00m^2. 4 Panes 3.00–4.00m^2. 5 Panes $\geq$ 4.00m^2: size stated.	1 Method of disposal of debris where not at the discretion of the contractor. 2 Materials to be kept for re-use.	

28 Floor, wall, ceiling and roof finishings

In-situ, tiled, block, Mosaic, sheet, applied liquid or planted finishes

Drawings that must accompany this section of measurement.	Mandatory information to be provided.	Notes, comments and glossary
1 Plans of each floor. 2 Principal sections through building. 3 External elevations. 4 Finishes schedules.	1 Kind, quality and size of materials. 2 Number of coats of in-situ work. 3 Types of underlays or linings. 4 Bedding or other method of fixing. 5 Type of pointing. 6 Laps. 7 Layout and width of joints and bays. 8 Patterns. 9 Spacing of battens and counter battens. 10 Composition and mix of mortars. 11 Bonding to other work. 12 Radius of curved work. 13 Nature of base. 14 Laid in one operation with base. 15 Applied finishes/sealers/polishes.	1 Area measured is net in contact with base. 2 Widths stated are the width of each finished face. 3 No deduction for voids $\leq 1.00m^2$. 4 No allowance to be made in net area for laps, dressings or any other labour. 5 All work is deemed internal unless stated as external.

Minimum information that must be shown on the drawings that accompany this section of measurement.		Works and materials deemed included.	
	1 All floor-to-floor or ceiling heights.		1 All rough and fair square, raking and curved cutting and waste.
	2 Layout of patterned work.		2 All ends and angles either formed or proprietary.
	3 Details of moulded work.		3 Extra material for curved work.
			4 Forming all rough and fair grooves, throats, mortices, chases, edges, rebates, holes, stops, mitres and the like labours.
			5 Raking out joints to form a key.
			6 Raking out joints to insert skirtings or the like.
			7 All work in forming returns, ends, internal and external angles.
			8 All extra material required for bonding.
			9 Additional material in laps and dressings.
			10 All work on attached columns and beams
			11 Work in forming voids and holes ≤ 1.00 m².
			12 Formwork or any other form of temporary support.
			13 Working finishes up to and around all accessories.

Item or work to be measured	Unit	Level one	Level two	Level three	Notes, comments and glossary
1 Screeds, beds and toppings, thickness and number of coats stated	m m²	1 ≤ 600mm wide. 2 > 600mm wide.	1 Level and to falls only ≤ 15° from horizontal. 2 To falls, cross falls and slopes ≤ 15° from horizontal. 3 To falls, cross falls and slopes > 15° from horizontal.	1 Surface finish. 2 Nature of background. 3 Backings and beddings: thickness stated. 4 Underlays: type and thickness stated. 5 Insulation: type and thickness stated. 6 Laid in one operation with its base. 7 Pattern of joints. 8 Laid in one operation with base (monolithically).	1 Thicknesses stated are the thickness exclusive of adhesives, keys, grooves and the like. 2 Underlays include plasterboard or other rigid sheet lathing.
2 Finish to floors, type of finish and overall thickness stated	m m²	1 ≤ 600mm wide. 2 > 600mm wide.			3 Tiles are deemed to be laid with their long edges vertical or parallel to the long axis of the floor or ceiling.
3 Raised access floors, type of finish and thickness of panels stated	m²		1 Height of cavity stated.		
4 Ramps to raised access floors	nr	1 Length, width and height.			4 The height stated is the height at the top end.
5 Fire barriers within void below raised floor	m² m	1 Thickness stated. 2 Thickness and height stated.	1 Fire rating. 2 Method of fixing in position. 3 Obstructed by services.		1 Fire barriers are deemed to include all support work, scribing or forming to fit, angles, ends and working around structures, support work and services.

Item or work to be measured	Unit	Level one	Level two	Level three	Notes, comments and glossary
6 Finish to roofs, type of finish and overall thickness stated	m, m²	1 ≤ 600mm wide. 2 > 600mm wide.	1 Level and to falls only ≤ 15° from horizontal. 2 To falls, cross falls and slopes ≤ 15° from horizontal. 3 To falls, cross falls and slopes > 15° from horizontal.		1 Roof finishes include grass, sedum and the like live finishes. 2 They exclude solar reflective paint, chippings and the like that will be measured with their associated roof covering. 3 Width is the width of each face.
7 Finish to walls, type of finish and overall thickness stated	m, m²	1 ≤ 600mm wide. 2 > 600mm wide.	1 Curved, radius stated.		1 Height of wall finishes is measured to finished ceiling height or specified height past suspended ceilings. 2 Wall finishes are measured behind skirtings unless the skirting is installed prior to the wall finish. 3 Width is the width of each face.
8 Finish to isolated columns, type of finish and overall thickness stated					
9 Finish to ceilings, type of finish and overall thickness stated	m, m²	1 ≤ 600mm wide. 2 > 600mm wide.	1 Over 3.50m above structural floor level.		1 Height of ceiling is measured from structural floor level to soffit of finished ceiling or isolated beam. 2 Width is the width of each face.
10 Finish to isolated beams, type of finish and overall thickness stated					
11 Finish to treads	m	1 Net width stated.	1 Curved: radius stated.	1 Nature of background. 2 Insets described.	
12 Finish to risers	m	1 Net width stated.	1 Curved: radius stated. 2 Undercut.	1 Nature of background.	
13 Finish to strings and aprons	m	1 Net width or girth on face stated.	1 Raking. 2 Sloping. 3 Curved: radius stated.	1 Nature of background.	
14 Skirtings, net height stated	m	1 Net height stated.	1 Raking. 2 Sloping. 3 Curved: radius stated.	1 Nature of background. 2 Cove formers.	

Item or work to be measured	Unit	Level one	Level two	Level three	Notes, comments and glossary
15 Linings to channels	m	1 Net girth on face stated.	1 Raking. 2 Sloping. 3 Curved: radius stated.	1 Nature of background.	1 Ends, angles and outlets deemed included.
16 Kerbs and cappings	m	1 Net girth on face stated.			1 Measure the length in contact with the base. 2 All ends, angles, intersections are deemed included.
17 Coves 18 Mouldings 19 Cornices 20 Architraves 21 Ceiling ribs 22 Bands	m	1 Girth and shape stated. 2 Dimensioned description.	1 Patterned: details stated. 2 Horizontal. 3 Raking. 4 Sloping. 5 Vertical. 6 Undercut. 7 Flush. 8 Raised. 9 Sunk.	1 Method of fixing. 2 Nature of background.	
23 Special tiles, slabs or blocks	nr - - - - m²	1 Dimensioned description. 2 Manufacturers' reference.			1 Special tiles are only measured here if not already measured under previous rules 11–22 inclusive.
24 Surface dressings, sealers or polishes.	m²	1 Type stated. 2 Rate of coverage stated.	1 Horizontal. 2 Sloping. 3 Vertical. 4 Soffits.	1 Nature of finish being treated.	1 Dressings include carborundum grains, stone chippings and the like. 2 Sealers include waterproofers, hardeners, dustproofers, polishes and the like.
25 Movement joints 26 Cover strips 27 Dividing strips 28 Beads, function stated 29 Nosings	m	1 Dimensioned description.	1 Depth of suspension.	1 Method of fixing. 2 Nature of background.	1 Movement joints include expansion joints. 2 Function of beads include angle beads, stop beads, shadow gap beads, casing beads and the like.
30 Reinforcement, details stated 31 Metal mesh lathing, details stated 32 Board insulation, thickness stated 33 Quilt insulation, thickness stated 34 Isolation membranes, thickness stated	m - - - - m²	1 To walls. 2 To ceilings. 3 To floors. 4 To roofs. 5 To isolated beams. 6 To isolated columns.		1 Method of fixing. 2 Nature of background.	

Item or work to be measured	Unit	Level one	Level two	Level three	Notes, comments and glossary
35 Accessories	nr	1 Dimensioned description.	1 Method of fixing.	1 Nature of background.	1 Accessories include access panels, special panels in raised access floors, vent grilles, ornaments and the like.
36 Precast plaster components	m	1 Dimensioned description or proprietary reference.	1 Method of fixing.	1 Nature of background.	1 The dimensioned description shall fully describe the component including stating the end use such as cornice, moulded band, architrave and the like.

29 Decoration

Painting and clear finishes
Intumescent coatings
Decorative papers or fabrics
Anti corrosion treatments

		Mandatory information to be provided.	Notes, comments and glossary
Drawings that must accompany this section of measurement.	1 Plans and elevation drawings showing scope and location of the work. 2 Painting/decorating schedule(s).	1 Kind and quality of materials. 2 Nature of base. 3 Description of surface if not smooth. 4 Preparatory work to base. 5 Type and number of priming or sealing coats. 6 Type and number of undercoats. 7 Type and number of finishing coats. 8 Method of application. 9 Type of treatment applied between coats. 10 Pattern of decorative papers. 11 Method of fixing and jointing papers or fabrics.	1 Method of application left to contractor unless stated otherwise. 2 Nature of non-smooth base includes description of base texture and profile such as corrugated, fluted, moulded or carved.
Minimum information that must be shown on the drawings that accompany this section of measurement.	1 Ceilings over 3.5m above finished floor level.	1 Works and materials deemed included.	1 All cutting to line. 2 Multi-coloured work on differing surfaces. 3 All cutting to papers and fabrics.

Item or work to be measured	Unit	Level one	Level two	Level three	Notes, comments and glossary
1 Painting to general surfaces	m	1 ≤ 300mm girth.	1 Internal.	1 Work to ceilings or beams over 3.5m but not exceeding 5m above finished floor level and thereafter in 3.00m stages.	1 The area or girth measured is that covered.
2 Painting to glazed surfaces irrespective of pane sizes	m²	2 > 300mm girth.	2 External.	2 Surfaces to remain unpainted.	2 The girth of frames and the like is calculated from one edge to the other over all trims, architraves, stops and the like and assume doors have been removed prior to painting.
3 Painting structural metalwork	nr	3 Isolated areas ≤ 1.00m² irrespective of location or girth.		3 Multi-coloured on one surface.	3 The girth designated as external on door frames and the like is that part of the frame visible when the door is closed.
4 Painting radiators, type stated				4 Patterned: details stated.	
5 Painting gutters				5 Fire rating.	4 Examples of radiator types are flat, panelled, column, tubular or the like.
6 Painting pipes				6 Application on site prior to fixing.	
7 Painting services, type stated				7 Application off site prior to fixing.	
8 Painting railings, fences and gates	m / m² / nr	1 ≤ 300mm girth. / 2 > 300mm girth. / 3 Isolated areas ≤ 1.00m² irrespective of location or girth.	1 Closed. / 2 Open. / 3 Ornamental.		1 Closed means no gaps whatsoever.

Decorative papers or fabrics

Item or work to be measured	Unit	Level one	Level two	Level three	Notes, comments and glossary
9 Walls and columns	m	1 Areas ≤ 1.00m².	1 Curved surfaces. radii stated.	1 Lining paper.	1 No deduction made for voids ≤ 1.00m².
10 Ceilings and beams	m²	2 Areas > 1.00m².		2 Work to ceilings or beams over 3.5m but not exceeding 5m above finished floor level and thereafter in 3.00m stages.	
11 Borders	m	1 Dimensioned description.			1 Mitres, ends angles, scribing or cutting deemed included.
12 Motifs	nr	1 Dimensioned description or diagram.			

30 Suspended ceilings

Demountable suspended ceilings

Solid suspended ceilings

Drawings that must accompany this section of measurement.	Mandatory information to be provided.		Notes, comments and glossary
1 Plans showing location of the work. 2 Reflected ceiling plans showing scope and complexity of the work.	1 Kind and quality of materials. 2 Size(s) of panels and strips. 3 Construction of suspension framing and systems. 4 Method of fixing. 5 Nature of background. 6 Nature of services located in the ceiling void. 7 Nature of integral services and fittings.		1 All work is deemed internal unless described as external. 2 Secondary steel or timber support work is measured elsewhere.
Minimum information that must be shown on the drawings that accompany this section of measurement.	Works and materials deemed included.		
1 Services located within the ceiling void. 2 Location of integral services and fittings.	1 Working over and around obstructions. 2 All cutting. 3 Forming openings and holes. 4 All additional support work including bridging for fittings. 5 All extra work required for work described as patterned.		1 Integral fittings are those designed to be incorporated into the ceiling structure.

Item or work to be measured	Unit	Level one	Level two	Level three	Notes, comments and glossary
1 Ceilings 2 Plenum ceilings 3 Beams 4 Bulkheads	m²	1 Depth of suspension ≤ 150mm. 2 Depth of suspension 150–500mm. 3 Depth thereafter in 500mm stages.	1 Type and thickness of lining. 2 Method of fixing lining to suspension. 3 Integral insulation. 4 Integral vapour barrier. 5 Height of work exceeding 3.50m above finished floor level in 1.50m stages.	1 Patterned: details stated. 2 Sloping. 3 Curved: radius and plane of curve stated. 4 Suspension obstructed by services. 5 Trims at regular intervals within area of ceiling: details stated.	1 Area measured is that on the exposed face. 2 Depth of suspension is measured from underside of main structural soffit or secondary support work to back surface of lining. 3 Height of work is measured from finished floor level to face of ceiling. 4 Where ceilings are suspended from sloping, curved or irregular structures an average depth of suspension shall be stated.
5 Isolated strips	m	1 Width ≤ 600mm. 2 Width thereafter in 300mm stages.			
6 Upstands	m	1 Height ≤ 600mm. 2 Height thereafter in 300mm stages.			
7 Access panels	nr	1 Dimensioned description. 2 Proprietary reference.	1 Composition. 2 Method of fixing.	1 Method of locking.	1 Access panels are deemed to included all additional support work, framing, edge trim and fixings.
8 Edge trims 9 Angle trims	m	1 Dimensioned description.	1 Plain. 2 Floating.	1 Centres of fixing. 2 Nature of background.	1 Plain trims are those fixed to the structure. 2 Floating trims are those fixed to the ceiling system. 3 Trims are deemed to include ends and angles.
10 Fire barriers	m --- m²	1 Dimensioned description. --- 2 Thickness stated.	1 Fire rating where required.	1 Method of fixing in position. 2 Obstructed by services stated.	1 Fire barriers are deemed to include all support work, scribing or forming to fit, angles, ends and working around structures, support work and services.

Item or work to be measured	Unit	Level one	Level two	Level three	Notes, comments and glossary
11 Collars for services passing through fire barriers	nr	1 Diameter of pipe. 2 Size of trunking.	1 Pipes. 2 Trunking.	1 Length of sleeve each side of barrier stated. 2 Fire rating	1 Collars are only measured here where they are integral with the fire barrier. If they are not integral they shall be measured in accordance with the rules of Work Section 40: Builder's work in connection with mechanical, electrical and transportation installations.
12 Fittings	nr - - - - - m	1 Dimensioned description.		1 Nature of background stated.	
13 Shadow gap battens	m	1 Dimensioned description.		1 Centres of fixing. 2 Nature of background.	1 Ends, angles and the like deemed included.

31 Insulation, fire stopping and fire protection

Board, sheet, quilt, sprayed, loose fill or foamed insulation and fire protection installations

Drawings that must accompany this section of measurement.	1 Plans, sections and details sufficient to show the scope and location of the various works relating to this section.	**Mandatory information to be provided.** 1 Type, quality and thickness of materials. 2 Fire rating where required. 3 Extent of laps. 4 Method of fixing, laying or applying where not at the discretion of the contractor.	Notes, comments and glossary
Minimum information that must be shown on the drawings that accompany this section of measurement.		**Works and materials deemed included.** 1 All cutting. 2 Working around or over all members and services.	

Item or work to be measured	Unit	Level one	Level two	Level three	Notes, comments and glossary
1 Boards 2 Sheets 3 Quilts 4 Loose	m²	1 Thickness stated.	1 Plain areas. 2 Laid across joists, rafters, partition framing or like members: centres of members stated. 3 Laid between joists, rafters, partition framing or like members: centres of members stated.	1 Horizontal. 2 Vertical. 3 Sloping. 4 Soffit.	1 The area measured is that covered. 2 No deduction is made for voids ≤ 1.00m². 3 The area of joists, rafters, partition framing or like members is deducted when the material is laid between such members. 4 Sloping is the upper surface not horizontal. 5 Soffits are the underside of any horizontal or sloping structure.
	m	2 Purpose and dimensioned description.	4 Location stated.		
5 Sprayed	m²	1 Thickness stated.		1 Nature of background.	

Item or work to be measured	Unit	Level one	Level two	Level three	Notes, comments and glossary
6 Filling cavities	m²	1 Thickness stated.	1 Mineral fibre. 2 Plastic beads. 3 Cellulose fibre. 4 Expanding foam. 5 Other type of blown or injected material.	1 Internal. 2 External.	1 Drilling holes in structure to allow injection of material and subsequent making good is deemed included.
7 Fire stops, type stated	m	1 Dimensioned description. 2 Fire rating.	1 Horizontal. 2 Vertical. 3 Raking. 4 Stepped. 5 Curved, radius stated.	1 Method of fixing. 2 Background when mechanically fixed.	
8 Fire sleeves, collars and the like	nr	1 Dimensioned description. 2 Fire rating.		1 Method of fixing. 2 Background when mechanically fixed.	

32 Furniture, fittings and equipment

General fixtures, furnishings and equipment
Kitchen fittings
Catering equipment
Sanitary appliances and fittings
Notices and signs
Site and street furniture
Bird/vermin control

		Mandatory information to be provided.	Notes, comments and glossary
Drawings that must accompany this section of measurement.	1 Location drawings.		1 Sufficient information to design, procure or manufacture the item. 2 Kind and quality of materials. 3 Details of all associated building work. 4 Tests with which materials and equipment must comply.
Minimum information that must be shown on the drawings that accompany this section of measurement.	1 Components.	1 Works and materials deemed included.	1 Marking positions. 2 Connecting to services. 3 Commissioning. 4 All associated excavations and concrete foundations or bases unless stated as measured elsewhere.

Item or work to be measured	Unit	Level one	Level two	Level three	Notes, comments and glossary
1 Fixtures, fittings or equipment without services 2 Fixtures, fittings or equipment with services	nr	1 Component drawing reference. 2 Dimensioned diagram. 3 Manufacturers reference.	1 Ancillary items provided with the equipment. 2 Integral controls and indicators. 3 Remote controls and indicators. 4 Supports, mountings and brackets.	1 Method of fixing. 2 Nature of background.	1 Accepting delivery, unloading, transporting about site, storing and handling and disposing of all packing materials is deemed included.
3 Ancillary items not provided with the item of equipment	nr	1 Type, size and method of jointing stated.			
4 Fixtures, fittings or equipment supplied by the employer	nr	1 Type and size.	1 Collecting from location off site: details stated.		
5 Signwriting	nr	1 Dimensioned description.	1 Font.	1 Nature of background.	

33 Drainage above ground

Rainwater installations
Foul drainage installations

Drawings that must accompany this section of measurement.			Mandatory information to be provided.	1 Location of installation. 2 Nature of background. 3 Method and spacing of fixing. 4 Method and spacing of joints. 5 Description of material. 6 Type of brackets or supports. 7 Finish.	Notes, comments and glossary
1 Roof plan(s). 2 Floor plans of any floors that have drainage installations. 3 Principal sections and elevations.					
Minimum information that must be shown on the drawings that accompany this section of measurement.			Works and materials deemed included.	1 All joints. 2 All brackets and supports.	
1 Scope and location of work.					
Item or work to be measured	**Unit**	**Level one**	**Level two**	**Level three**	**Notes, comments and glossary**
1 Pipework	m	1 Nominal diameter.	1 Straight, curved, flexible. 2 Extendable. 3 Method of jointing.	1 Method of fixing to background.	1 Measured over all fittings.
2 Pipework ancillaries	nr	1 Dimensioned description.			1 Ancillaries include valves, non-return flaps, reducers, tapers, bends, hoppers, gullies, tun dishes, rodding eyes, traps, access doors, angles, offsets, shoes, sockets, tappings, bosses and the like. 2 The list is not exhaustive.
3 Items extra over the pipe in which they occur	nr	1 Fittings, pipe size ≤ 65mm. 2 Fittings, nominal pope size > 65mm.	1 One end. 2 Two ends. 3 Three ends. 4 Other, details stated.	1 With inspection door. 2 Method of jointing stated where different from pipe in which the fitting occurs.	1 Cutting and jointing pipes to fittings is deemed to be included. 2 Fittings which are reducing are measured extra over the largest pipe in which they occur.

Item or work to be measured	Unit	Level one	Level two	Level three	Notes, comments and glossary
4 Pipe sleeves through walls, floors and ceilings	nr	1 Type and nominal diameter of pipe. 2 Length of sleeve or thickness of structure being passed through. 3 Fire rating.	1 Type of structure being passed through described.	1 Method of fixing and packing. 2 Handing to others for fixing.	1 All making good including fire stopping is deemed included.
5 Gutters	m	1 Nominal size.	1 Straight, curved, flexible. 2 Extendable. 3 Method of jointing.	1 Method of fixing to background.	1 Measured over all fittings.
6 Gutter ancillaries	nr	1 Dimensioned description.			1 Ancillaries include stop ends, angles, outlets, overflows, tapers, reducers and the like. 2 The list is not exhaustive
7 Items extra over the gutter in which they occur	nr	1 Dimensioned description.			
8 Marking, position of and leaving or forming all holes, mortices, chases and the like required in the structure	item	1 Number and type of installations.			
9 Identification	nr	1 Plates. 2 Discs. 3 Labels. 4 Tapes. 5 Symbols or numbers. 6 Bands. 7 Charts or diagrams. 8 Other: type stated.	1 Details stated.	1 Method of fixing.	
10 Testing and commissioning	item	1 Installation stated.	1 Attendance required. 2 Insurances required by employer.	1 Preparatory operations stated. 2 Stage tests (nr) listed and purpose stated.	1 All fuel and power used is deemed included. 2 Provision of test certificates is deemed included.

Item or work to be measured	Unit	Level one	Level two	Level three	Notes, comments and glossary
11 Preparing drawings	item	1 Number of copies stated.	1 Method stated. 2 Information to be shown stated.	1 Working drawings. 2 As fitted drawings.	
12 Operating manuals and instructions	item				

34 Drainage below ground

Storm water drain systems
Foul drain systems
Pumped drain systems
Land drainage

Drawings that must accompany this section of measurement.		Mandatory information to be provided		Notes, comments and glossary
Minimum information that must be shown on the drawings that accompany this section of measurement.	1 Layout of drainage showing scope of work. 2 Invert depths. 3 Cover levels. 4 Pipe sizes. 5 Details of manholes, inspection chambers and other pits, tanks, chambers or the like. 6 Work outside site boundary.	Works and materials deemed included.	1 Drainage layout(s).	1 Kind and quality of materials.	1 Earthwork support. 2 Compacting bottoms of excavations. 3 Trimming excavations. 4 Backfilling with excavated materials. 5 Compacting backfill. 6 Disposal of surplus excavated materials. 7 Disposal of all water. 8 Those lengths of pipes within manhole walls. 9 Building in ends of pipes. 10 Bedding and pointing.	1 Work outside site boundary shall be measured separately. 2 Work below buildings shall be measured separately

Item or work to be measured	Unit	Level one	Level two	Level three	Notes, comments and glossary
1 Drain runs	m	1 Average trench depth in 500mm increments. 2 Type and nominal diameter of pipe. 3 Multiple pipes stating number and nominal diameter of pipes.	1 Method of jointing pipes. 2 Pipe bedding and or surround: details stated 3 Type of backfill if not obtained from the excavations.	1 Vertical. 2 Curved. 3 Below ground water level. 4 Next to existing roadway or path. 5 Next to existing building. 6 Specified multiple handling: details stated. 7 Disposal of excavated material where not at the discretion of the contractor: details stated.	1 Drain runs are measured from external face of manhole to external face of manhole or accessory. 2 Average depth is calculated for each run irrespective of maximum depth.
2 Items extra over drain runs irrespective of depth or pipe size	m²	1 Breaking up hard surface pavings: thickness stated. 2 Lifting and preserving turf: thickness stated.	1 Reinstating to match existing: details stated. 2 Relaying turf previously set aside.		1 The measurement of these extra over items shall be based on the designed width of beds in the trench. In the absence of a bed the width shall be calculated as the nominal size of the service plus 300mm subject to a minimum width of 500mm.
	m³	3 Breaking out hard materials. 4 Excavating in and removing hazardous material. 5 Excavating in running silt, running sand or other unstable ground. 6 Excavating below ground water level.	1 Details stated.		2 Hard material is any material which is of such a size, position or consistency that it can only be removed by special plant or explosives.
	m	7 Next to existing live services.			1 To be measured where precautions are specifically required. The method of protection is left to the discretion of the contractor.
	nr	8 Around existing live services crossing trench.			2 If in doubt the surveyor must measure an item giving the nature of the service.

Item or work to be measured	Unit	Level one	Level two	Level three	Notes, comments and glossary
3 Pipe fittings	nr	1 Dimensioned description.	1 Type stated. 2 Method of jointing to pipes.		1 Pipe fittings include bends, junctions, inspection pipes, rodding eyes and the like.
4 Accessories	nr				1 Accessories include gullies, traps, inspection shoes, fresh air inlets, non-return flaps, valves, rodding eyes and the like and are deemed to include gratings, covers, frames, baskets, filters and all other integral and or associated fittings. 2 The list is not exhaustive.
5 Pumps	nr	1 Size and capacity stated.	1 Integral controls. 2 Remote controls. 3 Indicators. 4 Supports.	1 Method of fixing. 2 Nature of background.	
6 Manholes 7 Inspection chambers 8 Soakaways 9 Cesspits 10 Septic tanks, 11 Other tanks and pits, type stated	nr	1 Detailed description stating maximum internal size of chamber. 2 Depth from top surface of cover to invert level in 250 mm stages. 3 Proprietary system, details stated.	1 Base slab thickness. 2 Wall thickness. 3 Cover slab dimensions. 4 Intermediate slab dimensions. 5 Benching dimensions. 6 Main channel diameter and configuration 7 Number and diameter of branch channels 8 Internal finish. 9 External finish.		1 Size stated is the maximum internal size of the chamber. 2 Rocker joints are deemed included. 3 Other tanks and pits would include catch pits, service chambers, stopcock pits and the like.

Item or work to be measured	Unit	Level one	Level two	Level three	Notes, comments and glossary
12 Extra over the excavation for:	m²	1 Breaking up hard surface pavings: thickness stated. 2 Lifting and preserving turf: thickness stated.			1 Making good around edges of surfaces is deemed included.
	m³	3 Breaking out hard materials. 4 Excavating in and removing hazardous material. 5 Excavating in running silt, running sand or other unstable ground. 6 Excavating below ground water level.			
13 Sundries	nr	1 Detailed dimensioned description.	1 Step irons. 2 Intercepting traps. 3 Backdrops. 4 Any other associated item: type stated.		1 Bedding, jointing and building in deemed included. 2 These items may be included with their associated manhole or chamber if part of a proprietary system.
14 Covers and frames	nr	1 Dimensioned description.	1 Manufacturers reference.	1 Method of fixing frame.	1 Bedding covers in grease is deemed included.
15 Marker posts	nr	1 Dimensioned description.	1 Identification plates: details stated.	1 Method of fixing: details stated.	1 Excavation, disposal of spoil and concrete for base deemed included.
16 Connections	item	1 Details stated.			
17 Testing and commissioning	item	1 Detailed description. 2 Type of test and standard to be achieved.	1 Attendance required. 2 Instruments and equipment to be provided.	3 Preparatory operations stated. 4 Stage tests (nr) listed.	1 Provision of all fuel, power, water and other supplies is deemed included. 2 Provision of test certificates is deemed included.

35 Site works

Road and path pavings
Hard landscaping
Sports surfacing

Drawings that must accompany this section of measurement.			Mandatory information to be provided.	1 Site plans and sections. 2 Component details.	Notes, comments and glossary 1 Unless stated as deemed included all associated excavations, disposals and fillings are measured in accordance with Work Section 5: Excavating and filling.
Minimum information that must be shown on the drawings that accompany this section of measurement.			Works and materials deemed included.	1 Scope and location of the work.	1 Kind, quality, shape and size of materials and components. 2 Treatment and layout of joints. 3 Nature of base. 4 Preparatory work. 5 Bedding and/or methods of fixing.
					1 All work is deemed external unless described as internal. 2 Formwork and all other temporary support. 3 Fair joints. 4 Fair edges. 5 Working over and around obstructions and into recesses and shaped inserts. 6 Forming shallow channels. 7 Cutting. 8 Disposal of surplus excavated material off site.			1 The areas measured are those in contact with the base. 2 No deductions made in superficial items for voids $\leq 1.00m^2$. 3 All thickness stated are nominal or finished after laying and compacting.

Item or work to be measured	Unit	Level one	Level two	Level three	Notes, comments and glossary
1 Kerbs 2 Edgings 3 Channels 4 Paving accessories	m	1 Dimensioned description.	1 Curved: radius stated.	1 Foundation: size and details stated. 2 Reinforcement: details stated. 3 Method of fixing or support.	1 Excavation and disposal is deemed included. 2 Ends, angles, outlets and intersections deemed included. 3 Edgings are deemed to include pegs and supports.

Item or work to be measured	Unit	Level one	Level two	Level three	Notes, comments and glossary
5 Extra over for:	nr	1 Special units. 2 Accessories.	1 Dimensioned description.		
6 In-situ concrete 7 Formwork 8 Reinforcement 9 Joints 10 Worked finishes 11 Accessories cast in		1 All measured in accordance with rules of Work Section 11: In-situ concrete works.			
12 Coated macadam and asphalt	m² - - - - m	1 Over 300mm wide: thickness stated. 2 Not exceeding 300mm wide: thickness and width stated.	1 Roads. 2 Pavings.	1 Level and to falls only. 2 To falls and crossfalls and slopes ≤ 15° from horizontal. 3 To slopes > 15° from horizontal. 4 Method of application. 5 Surface treatment or finish. 6 Special curing of finished work.	
	m	3 Thickness and girth on face stated.	3 Linings to channels.	7 Horizontal. 8 To falls.	1 Arrises, coves, ends, angles and outlets deemed included.
13 Gravel, hoggin and woodchip	m² - - - - m	1 Over 300mm wide: thickness stated. 2 Not exceeding 300mm wide: thickness and width stated.	1 Roads. 2 Pavings.	1 Level and to falls only. 2 To falls and crossfalls and slopes ≤ 15° from horizontal. 3 To slopes > 15° from horizontal. 4 Method of application. 5 Surface treatment or finish. 6 Special curing of finished work.	1 The area measured is that in contact with the base.

Item or work to be measured	Unit	Level one	Level two	Level three	Notes, comments and glossary
14 Interlocking brick and blocks, slabs, bricks, blocks, setts and cobbles	m² ---- m	1 Over 300mm wide. 2 Not exceeding 300mm wide: thickness and width stated.	1 Roads. 2 Pavings. 3 Treads. 4 Risers. 5 Margins or bands. 6 Channels.	1 Bedding: thickness stated. 2 Level and to falls only. 3 To falls, crossfalls and slopes ≤ 15° from horizontal. 4 To slopes > 15° from horizontal. 5 Laid in bays: average size of bays stated. 6 Joint pattern stated. 7 Curved: radius stated. 8 Foundation and haunching.	1 Setting pavings into recessed manhole covers is deemed included. 2 All cutting and fitting is deemed included. 3 Excavation and disposal is deemed included. 4 Ends, angles, outlets and intersections deemed included.
	m	3 Thickness and girth on face stated.	7 Linings to channels.	9 Horizontal. 10 To falls.	5 Arrises, coves, ends, angles and outlets deemed included.
15 Extra over for:	nr	1 Special units. 2 Accessories.	1 Dimensioned description.		1 Special units include gulley surrounds, hoppers and the like.
16 Accessories	m²	1 Thickness stated.	1 Membranes.		
	m	2 Dimensioned description.	2 Movement joints. 3 Expansion joints.		1 Joints located at the discretion of the contractor are not measurable.
17 Site furniture	nr	1 Dimensioned description. 2 Manufacturers reference.	1 Tree grilles. 2 Bollards. 3 Litter bins. 4 Seats. 5 Other: type stated.	1 Method of fixing.	
18 Liquid applied surfacings 19 Sheet surfacings 20 Tufted surfacings 21 Proprietary coloured sports surfacings 22 Surface dressings	m²	1 Thickness stated.	1 Number of coats or layers.	1 Level and to falls only. 2 To falls and crossfalls and slopes ≤ 15° from horizontal. 3 To slopes > 15° from horizontal. 4 Method of application or fixing. 5 Surface treatment or finish. 6 Special curing of finished work. 7 Extent of laps. 8 Type of seams.	

Item or work to be measured	Unit	Level one	Level two	Level three	Notes, comments and glossary
23 Line markings, width stated	m	1 Width ≤ 300mm. 2 Width > 300mm: width stated.	1 Straight, continuous. 2 Straight, broken. 3 Curved, continuous: radius stated. 4 Curved, broken: radius stated.	1 Number of coats. 2 Method of application. 3 Treatment applied between coats. 4 Colour.	
24 Letters, figures and symbols	nr	1 Dimensioned description.			
25 Sheet linings to pools, lakes, ponds, waterways and the like	m m²	1 Width ≤ 500mm 2 Width > 500mm	1 Horizontal. 2 Sloping, slope stated. 3 Curved, radius stated. 4 Vertical.	1 Underlays.	1 The area measured is net and is that in contact with base. 2 No deduction is made for voids ≤ 1.00m². 3 No allowance is made for laps and the like. 4 All other requirements of Work Section 19: Waterproofing will apply to the measurement of this work.
26 Spot items	nr	1 Dimensioned diagram. or 2 Dimensioned description.	1 Sumps. 2 Outlets. 3 Collars or sleeves. 4 Other; type stated.		1 Spot item work is deemed to include joints, dressing and bonding to surrounding work, undercloaks and other associated linings, ends, angles, bedding, pointing, fixings and the like.

36 Fencing

Drawings that must accompany this section of measurement.		1 Site plans and sections. 2 Component details.	**Mandatory information to be provided.**	1 Kind and quality of materials. 2 Method of construction. 3 Surface treatments or finishes applied as part of the manufacturing process or applied before delivery to site. 4 Nature and size of backfilling.	**Notes, comments and glossary** 1 Unless stated as deemed included all associated excavations, disposals and fillings are measured in accordance with Work Section 5: Excavating and filling.
Minimum information that must be shown on the drawings that accompany this section of measurement.		1 Scope and location of the work. 2 Location of fencing designed for sloping ground. 3 Pre-contract ground water level(s) and date(s) established.	**Works and materials deemed included.**	1 All excavations, backfilling and disposal off site of surplus materials. 2 Earthwork support. 3 Disposal of ground and surface water. 4 in-situ concrete for post bases and the like 5 Temporary supports. 6 Formwork.	
Item or work to be measured	**Unit**	**Level one**	**Level two**	**Level three**	**Notes, comments and glossary**
1 Fencing, type stated	m	1 Height of fence.	1 Spacing, height and depth of supports.	1 Set to curve but straight between posts. 2 Curved: radius stated. 3 Ground sloping > 15° from horizontal. 4 Lengths ≤ 3m.	1 Fencing is measured over all supports. 2 The height of fencing is measured from the finished surface of the ground to the top of the infilling, or where there is no infilling, to the top wire or rail unless otherwise stated.

Item or work to be measured	Unit	Level one	Level two	Level three	Notes, comments and glossary
1 [Fencing: type stated—cont.]			1 Spacing, height and depth of supports.	1 Set to curve but straight between posts. 2 Curved: radius stated. 3 Ground sloping > 15° from horizontal. 4 Lengths ≤ 3m.	3 Supports are posts, struts or the like occurring at regular intervals. 4 Curved fencing is fencing curved between supports. 5 The depth of supports and special supports is the depth below the ground surface or other stated base. 6 The height of supports and special supports is the height above the ground surface or other stated base.
2 Extra over for special supports	nr	1 Size, height and depth stated.	1 End post. 2 Angle or corner post. 3 Integral gate post. 4 Straining post. 5 Other: type stated.	1 Method of fixing. 2 Background stated. 3 Details of backstays or struts stated.	1 Integral and independent gate posts are deemed to include slamming stops and hanging pins or fillets.
3 Independent gate posts	nr				
4 Items extra over fencing, supports and special supports and independent gate posts irrespective of type	m³	1 Excavating below ground water level. 2 Excavating in hazardous material: details stated.			1 If the post-contract ground water level differs from the pre-contract ground water level then the measurements are adjusted accordingly.
	m³ m²	3 Breaking out existing. 4 Breaking up existing hard pavings: thickness stated.	1 Rock. 2 Concrete. 3 Reinforced concrete. 4 Brickwork, blockwork or stonework. 5 Tarmacadam or asphalt.		1 Rock is any hard material which is of such size or location that it can only be removed by the use of wedges, rock hammers, special plant or explosives. 2 A boulder ≤ 5m³ in volume or one that can be lifted out in the bucket of an excavator will not constitute rock. 3 Degraded or friable rock that can be scraped out by the excavator bucket does not constitute rock. 4 Making good existing hard pavings is deemed included.

Item or work to be measured	Unit	Level one	Level two	Level three	Notes, comments and glossary
5 Gates	nr	1 Height and width stated.	1 Type stated.	1 Power supply	1 Gates are deemed to include stops, catches, independent stays, sounders, warning lights and all associated work.
6 Ironmongery	nr	1 Type of item, unit or set stated.	1 Kind and quality of materials and fixings. 2 Surface finish. 3 Constituent parts of the units or sets.	1 Method of fixing. 2 Nature of base.	1 Ironmongery is deemed to include fixing with screws or bolts to match and preparing the base to receive it.

37 Soft landscaping

		Mandatory information to be provided.	Notes, comments and glossary
Drawings that must accompany this section of measurement.	1 Site plans and sections.	1 Kind, quality, size and composition of materials.	1 Unless stated as deemed included all associated excavations, disposals and fillings are measured in accordance with Work Section 5: Excavating and filling.
	2 Component details.	2 Preparatory work.	
	3 Planting schedule.	3 Timing of operations.	
		4 Size and type of pits, holes and trenches, either excavated or formed.	
		5 Types of supports and ties.	
		6 Special filling materials.	
		7 Method of labelling.	
		8 Work on roofs and the like stated.	
		Works and materials deemed included.	
Minimum information that must be shown on the drawings that accompany this section of measurement.	1 Scope and location of the work.	1 All work is deemed external unless described as internal.	1 The areas measured are those in contact with the base.
		2 All excavations and backfilling.	2 No deductions made in superficial items for voids ≤ 1.00m².
		3 All necessary multiple handling of excavated material.	3 All thickness stated are nominal or finished after laying and compacting.
		4 Disposal of surplus excavated material off site.	
		5 Removal of stones and rubbish.	
		6 Watering.	
		7 Labelling.	

Item or work to be measured	Unit	Level one	Level two	Level three	Notes, comments and glossary
1 Cultivating	m²	1 Depth stated.	1 Method and degree of tilth.		
2 Surface applications	m²	1 Type and rate stated. 2 Method of application state.	1 Herbicides. 2 Weedkillers. 3 Peat. 4 Manure. 5 Compost. 6 Mulch. 7 Fertiliser. 8 Sand. 9 Soil ameliorates. 10 Other: details stated.	1 Before sowing or planting. 2 After planting. 3 Around individual plants. 4 To general areas. 5 To beds. 6 To planters. 7 To pots.	1 Working in is deemed included.
3 Seeding	m²	1 Rate stated.	1 Grass seed. 2 Cultivated plant seed. 3 Wild flower seed. 4 Other: type stated.		1 Raking in, harrowing and rolling is deemed included.
4 Turfing	m² - - - - - m	1 Type and thickness stated. - - - - - - - - - - 2 Type, width and thickness stated.			1 Cutting is deemed to include trimming edges.

Item or work to be measured	Unit	Level one	Level two	Level three	Notes, comments and glossary
5 Trees	nr	1 Botanical name.	1 BS size designation and root system stated. 2 Girth, height and clear stem and root system stated.	1 Planting in cultivated or grassed areas prepared by others to be stated including all necessary reinstatement. 2 Planting indoors to be stated. 3 Details of initial cut back stated. 4 Details of watering stated. 5 Backfill type stated if not the material arising from excavations.	1 Supports and ties deemed included. 2 BS size include standard, advanced nursery stock or semi-mature types. 3 Young nursery stock includes seedlings, transplants and whips.
6 Young nursery stock trees	nr		3 Height and root system stated.		
7 Shrubs	nr				
8 Hedge plants	nr - - - - m		4 Height stated. - - - - - - - 5 Height, spacing, number of rows and layout stated.		
9 Plants	nr - - - - m^2		6 Size stated. - - - - - - - 7 Size and number per m^2 stated.		
10 Bulbs, corms and tubers	nr - - - - m^2		8 Size stated. - - - - - - - 9 Size and weight per m^2 stated.		
11 Plant containers	nr	1 Dimensioned description.	1 Linings: type stated.	1 Method of fixing.	

Item or work to be measured	Unit	Level one	Level two	Level three	Notes, comments and glossary
12 Protection	m	1 Temporary fencing: type and duration stated.	1 Ultimate ownership: details stated.		1 Permanent fencing is measured in accordance with the rules of Work Section 36: Fencing.
	nr	2 Tree guards, dimensioned description.	2 Type stated.	1 Method of fixing stated.	
	nr	3 Wrappings: height of wrapping and girth of tree stated.	3 Chemical application stated.	2 Method of fixing stated.	
	nr	4 Anti-desiccant sprays, height and girth of tree or spread of plant stated.	4 Type stated.	3 Rate of spray stated.	
	item	5 General: details stated.	5 Duration.		
13 Maintenance	item	1 Details stated.			

38 Mechanical services

Drawings that must accompany this section of measurement.	Mandatory information to be provided.	Notes, comments and glossary
1 Floor and site plans detailing layout of systems. 2 System schematics. 3 Cross sections and elevations. 4 Detail layouts for plantrooms and principal services installation areas. 5 Specific installation details for equipment and fittings. 6 Equipment schedules.	1 Type, quality and size or thickness of materials. 2 Method of fixing. 3 Location. 4 Where the mandatory information is available, measurement of pipework and/or ductwork will be fully detailed with the measurement of fittings identified and measured separately, in accordance with Alternative 1. However, where the mandatory information is not available then pipework/ductwork and associated insulation and fire protection is to be measured inclusive of fittings.	
Minimum information that must be shown on the drawings that accompany this section of measurement.	Works and materials deemed included.	
	1 All work is deemed internal unless stated as external. 2 All waste. 3 Extra material for labour made fittings. 4 All couplings, brackets, supports, fixings and cast in channels. 5 All labelling, tagging, identification and charts. 6 Marking of all holes chases and mortices. 7 Earth connectors, straps and links.	

Item or work to be measured	Unit	Level one	Level two	Level three	Notes, comments and glossary
1 Primary equipment	nr	1 System	1 Location of installation – Roofs, high or low level in plantrooms, risers or service ducts, high or low level on floors.	1 Type, size, capacity, load, rating, special finishes, casings, in-fills, associated integrated or remote ancillaries, controls, indicators or components, anti-vibration mountings, acoustic performance/treatment and method of fixing.	1 Primary equipment is defined as plant or equipment from which a system originates, e.g. boiler, main storage tank, air handling unit, fan and the like. 2 Cross referenced to drawings and/or a full specification. 3 High and low level to be in accordance with National Health and Safety recommendations and National Working Rule Agreements.
	nr	2 Off-load and position primary equipment.			
	nr	3 Assembly of composite items of primary equipment.			
2 Terminal equipment and fittings	nr	1 System.	1 Location of installation – roofs, high or low level in plantrooms, risers or service ducts, high or low level on floors.	1 Type, size, capacity, load, rating, special finishes, casings, in-fills, associated integrated or remote ancillaries, controls, indicators or components, anti-vibration mountings, acoustic performance/treatment and method of fixing.	1 Terminal equipment and fittings are defined as any item to which a system is distributed to, e.g. heat emitters, grilles, fan coil units and the like.
3 Pipework	m	1 Material, finish, nominal diameter, method of jointing, fixings and background for fixing.	1 Location of installation – Roofs, high or low level in plantrooms, risers or service ducts, high or low level on floors, in trenches.		1 Generally all pipework to be measured inclusive of all fittings, i.e. unions, connectors, flanges, bends, tees, junctions, reducers, test points, bosses, sockets, tappings and the like.
		1 Unless measured separately, all fittings are deemed to be included.			

Item or work to be measured	Unit	Level one	Level two	Level three	Notes, comments and glossary
4 Alternative 1 – pipe fittings	nr	1 Type, material, finish, nominal diameter; method of jointing.			1 Unions, connectors, flanges, bends, tees, sets, junctions, reducers, test points, bosses, sockets, tappings and the like.
5 Pipe ancillaries	nr	1 Type, material, finish, nominal diameter; method of jointing.			1 Valves, strainers, expansion bellows, anchors, guides and rollers, gullies, outlets, rainwater heads, tundishes, traps, pipe sleeves, wall, floor and ceiling plates and the like.
6 Ventilation ducts		1 Material, finish, section shape, dimensions, method of jointing, fixings and background for fixing.	1 Location of installation – roofs, high or low level in plantrooms, risers or service ducts, high or low level on floors.		1 Generally all ductwork to be measured inclusive of all fittings, i.e. connectors, flanges, bends, tees, junctions, reducers, spigots, test holes and the like.
		1 Unless measured separately, all fittings are deemed to be included.			
7 Alternative 1 – duct fittings	nr	1 Type, material, finish, shape, dimensions, method of jointing.			1 Joints, flanges, bends, tees, sets, junctions, reducers, spigots, test holes and the like.
8 Duct ancillaries	nr	1 Type, material, finish, shape, dimensions, method of jointing.			1 Dampers, in-duct heater/cooling coils.
9 Insulation and fire protection	nr	1 Type, material, finish	1 Location of installation – roofs, high or low level in plantrooms, risers or service ducts, high or low level on floors, in trenches.	1 To equipment and fittings.	
	m	2 Thickness, material, finish, nominal diameter of pipe.		1 To pipework.	1 Generally all insulation to pipework to be measured inclusive of all fittings, i.e. unions, connectors, flanges, bends, tees, sets, junctions, reducers, test points, bosses, sockets, tappings and the like.

Item or work to be measured	Unit	Level one	Level two	Level three	Notes, comments and glossary
9 Insulation and fire protection—cont.		1 Unless measured separately, all fittings are deemed to be included.			
10 Alternative 1 – Insulation and fire protection to pipe fittings	nr	1 Type, material, finish, nominal diameter; method of jointing.			1 Unions, connectors, flanges, bends, tees, sets, junctions. reducers, test points, bosses, sockets, tappings and the like.
11 Insulation and fire protection to pipe ancillaries	nr	1 Type, material, finish, nominal diameter; method of jointing.			1 Valves, strainers, expansion bellows, gullies, outlets, tundishes, traps, pipe sleeves, and the like.
12 Insulation and fire protection to ventilation ducts	m²	1 Thickness, material, finish, duct dimensions.	1 Location of installation – Roofs, high or low level in plantrooms, risers or service ducts, high or low level on floors, in trenches.		1 Generally all insulation to ductwork to be measured inclusive of all fittings, i.e. connectors, flanges, bends, tees, sets, junctions, reducers, spigots, test holes and the like, on the external girth of the duct to be insulated.
		1 Unless measured separately, insulation and fire protection to all fittings is deemed to be included.			
13 Alternative 1 – Insulation and fire protection to duct fittings		1 Type, material, finish, nominal diameter; method of jointing.			1 Joints, flanges, bends, tees, sets, junctions, reducers, spigots, test holes and the like.
14 Insulation and fire protection to equipment	m²	1 Thickness, material, finish, nominal area, performance rating.			1 Insulation to equipment measured separately where not given in the description of the items to which it relates.

Item or work to be measured	Unit	Level one	Level two	Level three	Notes, comments and glossary
15 Fire stopping	nr	1 Material, location, size of opening to be stopped, size of service stopping to be installed around, depth and fire rating.			
16 Identification	item	1 Plates. 2 Discs. 3 Labels. 4 Tapes. 5 Symbols or numbers. 6 Bands. 7 Charts or diagrams. 8 Other: type stated.			
17 Testing	item				1 To include chemical treatment, purging, sterilisation, pressure testing.
18 Commissioning	item				1 To include all final balancing, calibration and setting to work.
19 System validation	item				
20 Operation and maintenance manuals	item				
21 Drawing preparation	item	1 Classification of drawings.			1 Installation, co-ordination, as installed, and record drawings.
22 Training	item				
23 Loose ancillaries	item	1 Keys, tools, spares, chemicals and the like.			
24 Post-completion services	item	1 Maintenance, adjustments, servicing and the like.			1 See glossary of terms for definition of completion.

39 Electrical services

Drawings that must accompany this section of measurement.	Mandatory information to be provided.	Notes, comments and glossary
1 Floor and site plans detailing layout of systems. 2 System schematics. 3 Cross sections and elevations. 4 Detail layouts for plantrooms and principal services installation areas. 5 Specific installation details for equipment and fittings. 6 Equipment schedules.	1 Kind, quality and size or thickness of materials. 2 Method of fixing. 3 Location.	

Minimum information that must be shown on the drawings that accompany this section of measurement.	Works and materials deemed included.	
	1 All work is deemed internal unless stated as external. 2 All waste. 3 Extra material for labour made fittings. 4 All brackets, supports, fixings and cast in channels. 5 All labelling, tagging, identification and charts. 6 Marking of all holes chases and mortices. 7 Earth connectors, straps and links.	1 Where the mandatory information is available, measurement of cable containment and busbars will be fully detailed with the measurement of fittings identified and measured separately, in accordance with Alternative 1. However, where the mandatory information is not available then cable containment and busbars are to be measured inclusive of fittings.

Item or work to be measured	Unit	Level one	Level two	Level three	Notes, comments and glossary
1 Primary equipment	nr	1 System.	1 Location of installation – roofs, high or low level in plantrooms, risers or service ducts, high or low level on floors.	1 Type, size, capacity, load, rating, associated integrated or remote ancillaries, controls, indicators or components and method of fixing.	1 Primary equipment is defined as plant or equipment from which a system originates, e.g. main switch board, main control box and the like. 2 High and low level to be in accordance with National Health and Safety recommendations and National Working Rule Agreements.

Item or work to be measured	Unit	Level one	Level two	Level three	Notes, comments and glossary
1 Primary equipment—cont.	nr	2 Off-load and position primary equipment.			
	nr	3 Assembly of composite items of primary equipment.			
2 Terminal equipment and Fittings	nr	1 System.	1 Type, size, capacity, load, rating, special finishes, casings, in-fills, associated integrated or remote ancillaries, controls, indicators or components and method of fixing.	1 Location of installation – Roofs, high or low level in plantrooms, risers or service ducts, high or low level on floors.	1 Terminal equipment and fittings are defined as any item to which a system is distributed to, e.g. luminaires, switches, actuators and the like.
3 Cable containment	m	1 Type, material, finish, section shape, dimensions, number of compartments, method of jointing, fixings and background for fixing. Unless measured separately, all fittings are deemed to be included.	1 Location of installation – roofs, high or low level in plantrooms, risers or service ducts, high or low level on floors, in trenches.		1 Generally all cable containment to be measured inclusive of all fittings, i.e. joint boxes, connectors, flanges, bends, tees, junctions, reducers, spigots, fire barriers and the like.
4 Alternative 1 – cable containment fittings	nr	1 Type, material, finish, shape, dimensions, method of jointing.			1 Joint boxes, connectors, flanges, bends, tees, sets, junctions, reducers, spigots, fire barriers and the like.
5 Cables	m	1 Type, rating, size, number of cores, material, armouring, sheathing, method of jointing, fixings and background for fixing.	1 Location of installation – roofs, high or low level in plantrooms, risers or service ducts, high or low level on floors, in trenches.		

Item or work to be measured	Unit	Level one	Level two	Level three	Notes, comments and glossary
6 Cable terminations and joints	nr	1 Type, rating, size, number of cores, material, armouring, sheathing.			1 Includes all pots, seals, glands, lugs, connector blocks and shrouds.
7 Final circuits	nr	1 Cable type, rating, size, material, sheathing, number and type of points.	1 Location of installation – roofs, high or low level in plantrooms, risers or service ducts, high or low level on floors.		1 Includes all containment not measured separately, junction boxes, terminations, pots, seals, glands, lugs, connector blocks and shrouds. 2 Includes fixing containment or cables in chases, surface or suspended from soffits.
8 Modular wiring systems	nr	1 Cable type, rating, size, material, sheathing, number and type of points.			1 Includes all containment not measured separately, junction boxes, terminations, pots, seals, glands, lugs, connector blocks and shrouds. 2 Includes fixing containment or cables in chases, surface or suspended from soffits.
9 Busbar	m	1 Type, rating, material, number of bars, method of jointing, fixings and background for fixing.	1 Location of installation – Roofs, high or low level in plantrooms, risers or service ducts, high or low level on floors, in trenches.		1 Generally all busbar to be measured inclusive of all fittings, i.e. connectors, flanges, bends, tees, sets, junctions, feeder units, tap-off boxes fire barriers and the like.
		1 Unless measured separately, all fittings are deemed to be included.			
10 Alternative 1 – busbar fittings	nr	1 Type, material, finish, shape, dimensions, method of jointing.			1 Connectors, flanges, bends, tees, sets, junctions, feeder units, tap-off boxes fire barriers and the like. 2 Joints, flanges, bends, tees, sets, junctions, reducers, spigots, test holes and the like.
11 Tapes	m	1 Type, rating, size, material, sheathing, method of jointing, fixings and background for fixing.	1 Location of installation – Roofs, high or low level in plantrooms, risers or service ducts, high or low level on floors, in trenches.		1 Includes all connections, joints, test clamps.

Item or work to be measured	Unit	Level one	Level two	Level three	Notes, comments and glossary
12 Electrodes, earth rods, air terminations, termination bars	nr	1 Type, rating, size, material, method of jointing, fixings and background for fixing.			
13 Fire stopping and other associated fire protect work	nr	1 Material, location, size of opening to be stopped, size of service stopping to be installed around, depth and fire rating.			
14 Identification	item	1 Plates. 2 Discs. 3 Labels. 4 Tapes. 5 Symbols or numbers. 6 Bands. 7 Charts or diagrams. 8 Other: type stated.			
15 Testing	item				
16 Commissioning	item				
17 System validation	item				
18 Operation and maintenance manuals	item				
19 Drawing preparation	item	1 Classification of drawings.			1 Installation, co-ordination, as installed and record drawings.
20 Training	item				
21 Loose ancillaries	item	1 Keys, tools, spares, chemicals.			
22 Post practical completion Services	item	1 Maintenance, adjustments and servicing.			

40 Transportation

Drawings that must accompany this section of measurement.		Mandatory information to be provided.	Notes, comments and glossary
	1 Location plans	Works and materials deemed included.	1 Kind, quality and size or thickness of materials. 2 Location. 3 Earth connectors, straps and links.
Minimum information that must be shown on the drawings that accompany this section of measurement.	1 Overall dimensions of items or systems		1 All work is deemed internal unless stated as external. 2 All waste. 3 Extra material for labour made fittings. 4 All brackets, supports, fixings and cast in channels. 5 All labelling, tagging, identification and charts. 6 Marking of all holes chases and mortices.

Item or work to be measured	Unit	Level one	Level two	Level three	Notes, comments and glossary
1 System	nr	1 Type, size, capacity, load, rating, length, number of stops, storey height, associated integrated or remote ancillaries, controls, indicators or components.			1 Cross referenced to drawings and/or a full specification detailing finishes.
	nr	2 Off-load and position primary equipment.			
	nr	3 Assembly of composite items of primary equipment.			
	item	4 Free issue fixing steelwork and other components for installation by others.	1 Type of free issue materials.		
	item	5 Interface with and connection to systems supplied and installed by others.	1 Type of system.	1 Nature of interface.	1 Systems supplied and installed by others will include telephones, intercoms, fire alarm and the like.

Item or work to be measured	Unit	Level one	Level two	Level three	Notes, comments and glossary
2 Fire stopping and other associated fire protect work	nr	1 Material, location, size of opening to be stopped, size of service stopping to be installed around, depth and fire rating.			
3 Identification	item	1 Plates. 2 Discs. 3 Labels. 4 Tapes. 5 Symbols or numbers. 6 Bands. 7 Charts or diagrams. 8 Other: type stated.			
4 Testing and commissioning	item				
5 System validation	item				
6 Operation and maintenance manuals	item				
7 Drawing preparation	item	1 Classification of drawings.			1 Installation, co-ordination, as installed and record drawings.
8 Training	item				
9 Loose ancillaries	item	1 Keys, tools, spares, chemicals.			
10 Post practical completion services	item	1 Maintenance, adjustments and servicing.			

41 Builder's work in connection with mechanical, electrical and transportation installations

		Mandatory information to be provided		Kind and quality of materials.	Notes, comments and glossary
Drawings that must accompany this section of measurement.				1 Layout of each type of service installation.	
Minimum information that must be shown on the drawings that accompany this section of measurement.				1 Construction of structure. 2 Fire compartmentation. 3 Service runs.	

Item or work to be measured	Unit	Level one	Level two	Level three	Notes, comments and glossary
Work for services installations in new buildings					
1 General builder's work in connection with:	item	1 Type of Installation stated.			1 An item shall be given for each and every services installation. 2 Examples of installations are cold water services, hot water services, lighting installations, power installations, lift installations. In the case of large projects it may be necessary to sub-divide each installation into locations. 3 This item means every type of general builder's work necessary in connection with the service installation except those items included in the following rules.
2 Marking position of holes, mortices and chases in the structure	item				
3 Pipe and duct sleeves	nr	1 Size and type stated.	1 Fire rating stated.	1 Nature and thickness of structure stated. 2 Method of fixing stated. 3 Sleeves supplied by others stated.	1 Making good around sleeve is deemed included.
4 Bases, plinths and the like	nr	1 Size stated.	1 Method of forming or construction stated.	1 Anti-vibration pads. 2 Acoustic pads.	

Item or work to be measured	Unit	Level one	Level two	Level three	Notes, comments and glossary
5 Duct covers and frames	m - - - - nr	1 Width and type stated. - - - - - - - - - - - 2 Size and type stated.	1 Limitations to lengths of covers stated. 2 Finish stated.	1 Method of fixing stated. 2 Nature of background stated.	
6 Supports for services not provided by services contractor	m - - - - nr	1 Size and type stated. - - - - - - - - - - - 2 Size, length and type stated.	1 Pylons. 2 Poles. 3 Wall brackets. 4 Soffit hangers. 5 Stays. 6 Proprietary supports. Manufacturers reference stated.		
7 Catenary cables	m	1 Size and type stated.	1 Eye bolts: details stated. 2 Shackles: details stated. 3 Straining screws: details stated.		1 The length is net with no allowance made for sag.
Work for services installations in existing buildings					
8 Cutting holes through existing structures	nr	1 Size stated.	1 Nature and thickness of structure stated.		1 Making good to match existing or preparing for new work is deemed included.
9 Cutting mortices and sinkings in existing structure	nr	1 Size stated.	1 Nature of structure stated.		

Item or work to be measured	Unit	Level one	Level two	Level three	Notes, comments and glossary
10 Cutting chases through existing structures	m	1 Size and number of services stated.			
11 Lifting and replacing floor boards	m	1 Size and number of services stated.			1 No distinction is made between routes parallel or at an angle to the floor boards.
12 Lifting and replacing duct covers or chequer plates	m	1 Width and type stated.			2 Cutting boards and notching or holing joists is deemed included.
Work for external services installations					
13 Underground service runs	m	1 Average trench depth in 500mm increments. 2 Type and nominal diameter of pipe duct. 3 Multiple pipe ducts stating number and nominal diameter of pipe ducts. 4 Pipe duct(s) supplied by others; type and nominal diameter. 5 Type and size of cover tile(s) or identification tape(s).	1 Method of jointing pipe ducts. 2 Pipe duct bedding and or surround: details stated. 3 Type of backfill if not obtained from the excavations.	1 Vertical. 2 Curved. 3 Below ground water level. 4 Next to existing roadway or path. 5 Next to existing building. 6 Specified multiple handling: details stated. 7 Disposal of excavated material where not at the discretion of the contractor: details stated.	1 Pipe duct runs are deemed to run straight unless stated otherwise. 2 Pipe duct runs are measured from external face of manhole to external face of manhole or accessory. 3 Average depth is calculated for each run irrespective of maximum depth.
14 Items extra over service runs irrespective of depth or pipe size	m²	1 Breaking up hard surface pavings: thickness stated. 2 Lifting and preserving turf: thickness stated.	1 Reinstating to match existing: details stated. 2 Relaying turf previously set aside.		1 The measurement of these extra over items shall be based on the designed width of beds in the trench. In the absence of a bed the width shall be calculated as the nominal size of the service plus 300mm subject to a minimum width of 500mm.

Item or work to be measured	Unit	Level one	Level two	Level three	Notes, comments and glossary
14 [Items extra over service runs irrespective of depth or pipe size—cont.]	m³	3 Breaking out hard materials. 4 Excavating in and removing hazardous material. 5 Excavating in running silt, running sand or other unstable ground. 6 Excavating below ground water level.	3 Details stated.		1 Hard material is any material which is of such a size, position or consistency that it can only be removed by special plant or explosives.
	m	7 Next to existing live services.			1 To be measured where precautions are specifically required. 2 The method of protection is left to the discretion of the contractor.
	nr	8 Around existing live services crossing trench.			
15 Pipe duct fittings	nr	1 Dimensioned description.	1 Type stated.	1 Method of jointing to pipe ducts.	1 Pipe duct fittings include bends, junctions, inspection pipes, stop ends and the like and connections to pipes.
16 Accessories	nr	1 Dimensioned description.			1 Accessories include all integral and or associated fittings and connections to pipes.
17 Manholes 18 Access chambers 19 Valve chambers 20 Inspection chambers 21 Surface boxes 22 Stopcock pits	nr	1 Detailed description stating maximum internal size of chamber. 2 Depth from top surface of cover to top surface of base or invert level in 250mm stages. 3 Proprietary chambers boxes and the like: details stated.	1 Base slab thickness. 2 Wall thickness. 3 Cover slab dimensions. 4 Intermediate slab dimensions. 5 Internal finish. 6 External finish. 7 Cover and frame: type stated.		1 Size stated is the maximum internal size of the chamber. 2 Rocker joints are deemed included.

TABULATED WORK SECTIONS

Item or work to be measured	Unit	Level one	Level two	Level three	Notes, comments and glossary
23 Extra over the excavation for:	m²	1 Breaking up hard surface pavings; thickness stated. 2 Lifting and preserving turf; thickness stated.			1 Making good around edges of surfaces is deemed included.
	m³	3 Breaking out hard materials. 4 Excavating in and removing hazardous material. 5 Excavating in running silt, running sand or other unstable ground. 6 Excavating below ground water level.			
24 Marker posts 25 Marker plates	nr	1 Dimensioned description.	1 Identification plates; details stated. 2 Lettering required.	1 Method of fixing; details stated.	1 Excavation, disposal of spoil and concrete for base deemed included.
26 Connections	item	1 Details stated.			
27 Testing and commissioning	item	1 Detailed description. 2 Type of test and standard to be achieved.	1 Attendance required. 2 Instruments and equipment to be provided.		1 Provision of water and other supplies is deemed included. 2 Provision of test certificates is deemed included.

Appendices

Appendix A: Guidance on the preparation of bill of quantities

A.1 Bill of quantities breakdown structure

A.1.1 There a number of different breakdown structures for bill of quantities (BQ). They each have their own advantages and disadvantages. However, computerised BQ production systems with multiple sort facilities can be used to generate different BQ formats and make it easy to trace items – as long as items have been properly codified (refer to 2.15 (Codification of bill of quantities).

A.1.2 *NRM 2: Detailed measurement for building works* can be adopted as the rules of measurement for building works irrespective of what BQ breakdown structure is chosen.

A.1.3 The main BQ breakdown structures are:

(a) **Elemental**: Measurement and description is done by group elements and each group element forms a separate section of the BQ, irrespective of the order of work sections in *NRM 2: Detailed measurement for building works*. Group elements are sub-divided through the use of elements, which are further sub-divided by sub-elements. The group elements, elements and sub-elements used are those defined by *NRM 1: Order of cost estimating and cost planning for capital building works*.

Figure A.1: Elemental BQ breakdown structures for a simple building project

Elemental breakdown structure	
Bill No. 1:	Preliminaries (main contract)
Bill No. 2:	Facilitating Works
Bill No. 3:	Substructure
Bill No. 4:	Superstructure
Bill No. 5:	Internal Finishes
Bill No. 6:	Fittings, Furnishings and Equipment
Bill No. 7:	Services
Bill No. 8:	External Works
Bill No. 9:	Risks
Bill No. 10:	Provisional Sums
Bill No. 11:	Credits
Bill No. 12:	Daywork (provisional)

An elemental breakdown structure supports a logical and structured approach to the quantification of a building project. Moreover, this type of breakdown structure makes it easier for the quantity surveyor/cost manager to analysis a contractor's tender price and collects real-time cost data for future use.

(b) **Work section**: Measurement and description is divided into the work sections listed in *NRM 2: Detailed measurement for building works* (refer to the tabulated rules in Part 3 of these rules).

Figure A.2: Work section BQ breakdown structures for a simple building project

Work section breakdown structure	
Bill No. 1:	Main contractor's preliminaries
Bill No. 2:	Off-site manufactured materials, components and buildings
Bill No. 3:	Demolitions
Bill No. 4:	Alterations, repairs and conservation
Bill No. 5:	Excavating and filling
Bill No. 6:	Ground remediation and soil stabilisation
Bill No. 7:	Piling
Bill No. 8:	Underpinning
Bill No. 9:	Diaphragm walls and embedded retaining walls
Bill No. 10:	Crib walls, gabions and reinforced earth
Bill No. 11:	In-situ concrete works
Bill No. 12:	Precast/composite concrete
Bill No. 13:	Precast concrete
Bill No. 14:	Masonry
Bill No. 15:	Structural metalwork
Bill No. 16:	Carpentry
Bill No. 17:	Sheet roof coverings
Bill No. 18:	Tile and slate roof and wall coverings
Bill No. 19:	Waterproofing
Bill No. 20:	Proprietary linings and partitions
Bill No. 21:	Cladding and covering
Bill No. 22:	General joinery
Bill No. 23:	Windows, screens and lights
Bill No. 24:	Doors, shutters and hatches
Bill No. 25:	Stairs, walkways and balustrades
Bill No. 26:	Metalwork
Bill No. 27:	Glazing
Bill No. 28:	Floor, wall, ceiling and roof finishings
Bill No. 29:	Decoration
Bill No. 30:	Suspended ceilings
Bill No. 31:	Insulation, fire stopping and fire protection
Bill No. 32:	Furniture, fittings and equipment
Bill No. 33:	Drainage above ground
Bill No. 34:	Drainage below ground
Bill No. 35:	Site works
Bill No. 36:	Fencing
Bill No. 37:	Soft landscaping
Bill No. 38:	Mechanical services
Bill No. 39:	Electrical services
Bill No. 40:	Transport
Bill No. 41:	Builder's work in connection with mechanical, electrical and transportation installations
Bill No. 42:	Risks
Bill No. 43:	Provisional sums
Bill No. 44:	Credits
Bill No. 45:	Daywork (Provisional)

This breakdown structure is often preferred by contractors for the purpose of pricing as all like products and components are grouped together (e.g. the reinforced concrete columns, beams, floors, roofs and staircases), whereas they can be spread among a number of different elements when an elemental breakdown structure is used. Codification of BQ items using computerised BQ systems will solve the problem of preferences (refer to 2.15 (Codification of bill of quantities) of the rules).

(c) **Work package**: Measurement and description is divided into employer, quantity surveyor or contractor defined work packages – whichever is applicable. Works packages can be based on either a specific-trade (e.g. concrete work, brickwork and blockwork, roof coverings, painting and decorating, and wall tiling) or a single package comprising a number of different trades (e.g. a groundworks package might include all excavation and earthworks, below ground drainage and the ground-bearing concrete floor-slab – so as to make a single works contractor responsible for the interface between the drainage and the ground-bearing concrete floor-slab).

Figure A.3: Typical BQ breakdown structure for discrete work package

Work section breakdown structure	
Bill No. 1:	Main contractor's preliminaries
Bill No. 2:	Intrusive investigations
Bill No. 3:	Demolition works
Bill No. 4:	Groundworks
Bill No. 5:	Piling
Bill No. 6:	Concrete works
Bill No. 7:	Roof coverings and roof drainage
Bill No. 8:	External and internal structural walls
Bill No. 9:	Cladding
Bill No. 10:	Windows and external doors
Bill No. 11:	Mastic
Bill No. 12:	Non-structural walls and partitions
Bill No. 13:	Joinery
Bill No. 14:	Suspended ceilings
Bill No. 15:	Architectural metalwork
Bill No. 16:	Tiling
Bill No. 17:	Painting and decorating
Bill No. 18:	Floor coverings
Bill No. 19:	Fittings, furnishings and equipment
Bill No. 20:	Combined mechanical and electrical engineering services
Bill No. 21:	Lifts and escalators
Bill No. 22:	Facade access equipment
Bill No. 23:	External works and drainage
Bill No. 24:	Risks
Bill No. 25:	Provisional sums
Bill No. 26:	Credits
Bill No. 27:	Daywork (Provisional)

Again, codification of BQ items using computerised BQ systems will allow the re-sorting of items from elements to works packages for the purposes of tendering, and vice versa for the purpose of overall cost control (refer to 2.15 (Codification of bill of quantities) at Part 2 of these rules).

This breakdown structure is usually used by contractors to procure packages of work from their supply chain.

A.2 Bill of quantities breakdown structure for projects comprising more than one building

Where a building project comprises more than one type of building, it is recommended that a separate bill of quantities be prepared for each building; culminating in a 'summary bill' for the entire building project.

A.3 Order of items in bill of quantities

The order of items in a bill of quantities (BQ) is:

(1)   For elemental BQ:

(a)   Elements as contained in *NRM 1: Order of cost estimating and cost planning for capital building works.*

(b)   Within each element the order of measured items is cubic, square, linear, enumerated items and itemised items.

(c)   Labour-only items are to precede labour and material items within the subdivisions in (b).

(d)   Items within each subdivision in (b) and (c) above are to be placed in order of value, least expensive first.

(e)   Preambles are to be incorporated in each element as appropriate.

(f)   PC sums are to be incorporated in item descriptions.

(g)   Contractor designed works are to be incorporated under the applicable element or sub-element, after measured work, under a heading of 'Contractor designed work'. A price analysis for contractor designed work is to be incorporated (see paragraph 2.9.2 (Contractor designed work) in Part 2 (Rules for detailed measurement of building works) of these rules).

(h)   Provisional sums are to be listed and described in a separate bill.

(i)   Construction risks to be transferred to the contractor are to be identified and described in a 'schedule of construction risks' (see paragraph 2.10.2 (Risk transfer to the contractor) in Part 2 (Rules for detailed measurement of building works) of these rules).

(j)   Where applicable, provision for the contractor to offer credits against items and components arising from demolition works and/or soft strip works is to be provided (see paragraph 2.12 (Credits) in Part 2 (Rules for detailed measurement of building works) of these rules).

(2)   For work section BQ:

(a)   Work sections as contained in *NRM 2: Detailed measurement for building works*, although separate locational BQ sections such as facilitating works, substructure, superstructure and or external works might be required.

(b)   Subdivisions:

(i)   of work sections as contained in *NRM 2: Detailed measurement for building works*;

(ii)   as required by *NRM 2: Detailed measurement for building works*, such as external paintwork;

(iii)   of different types of materials; such as different mixes of concrete, different types of brick.

(c)   Within each subdivision in (b), the order of cubic, square, linear, enumerated items and itemised items.

(d)   Labour-only items are to precede labour and material items within the subdivisions in (c).

(e)   Items within each subdivision in (c) and (d) above are to be placed in order of value, least expensive first.

(f)   Preambles are to be incorporated in the appropriate work section.

(g)   PC sums are to be incorporated in item descriptions.

(h)   Contractor designed works are to be incorporated under the applicable element or sub-element, after measured work, under a heading of 'Contractor designed work'. A price analysis for contractor designed work is to be incorporated (see paragraph 2.9.2 (Contractor designed work) in Part 2 (Rules for detailed measurement of building works) of these rules).

(i) Provisional sums are to be listed and described in a separate bill.

(j) Construction risks to be transferred to the contractor are to be identified and described in a 'schedule of construction risks' (see paragraph 2.10.2 (Risk transfer to the contractor) in Part 2 (Rules for detailed measurement of building works) of these rules).

(k) Where applicable, provision for the contractor to offer credits against items and components arising from demolition works and/or soft strip works is to be provided (see paragraph 2.12 (Credits) in Part 2 (Rules for detailed measurement of building works) of these rules).

(3)   For work package BQ:

(a) Work packages as defined by the employer, quantity surveyor or contractor, whichever is applicable.

(b) Within each work package the order of measured items is cubic, square, linear, enumerated items and itemised items.

(c) Labour-only items are to precede labour and material items within the subdivisions in (b).

(d) Items within each subdivision in (b) and (c) above are to be placed in order of value, least expensive first.

(e) Preambles are to be incorporated in the appropriate work package.

(f) PC sums are to be incorporated in item descriptions.

(g) Contractor designed works are to be incorporated into the applicable work package bill, after measured work, under a heading of 'Contractor designed work', under a heading of 'Contractor designed work'. A price analysis for contractor designed work is to be included (see paragraph 2.9.2 (Contractor designed work) in Part 2 (Rules for detailed measurement of building works) of these rules).

(h) Provisional sums are to be listed and described in a separate bill.

(i) Construction risks to be transferred to the contractor are to be identified and described in a 'schedule of construction risks' (see paragraph 2.10.2 (Risk transfer to the contractor) in Part 2 (Rules for detailed measurement of building works) of these rules).

(j) Where applicable, provision for the contractor to offer credits against items and components arising from demolition works and/or soft strip works is to be provided (see paragraph 2.12 (Credits) in Part 2 (Rules for detailed measurement of building works) of these rules).

A.4　Format of bill

The bill for each element or work section is to be commenced on a new sheet. The ruling of the paper and typical headings for each type of bill are shown in Figures A.4 and A.5.

Figure A.4: Typical BQ format for an elemental bill of quantities

Bill No. 3 Superstructure							
3.2.5 External walls							
3.2.5.1	EXTERNAL WALLS ABOVE GROUND FLOOR LEVEL						
3.2.5.1.1	Common brickwork in cement: lime mortar (1:1:6)						
3.2.5.1.1.1	Walls 102.5mm thick, brickwork; built against other work	196	m²				
3.2.5.1.1.2	Walls 215mm thick, brickwork	302	m²				

Figure A.5: Typical BQ format for a work section bill of quantities

Bill No. 2 Superstructure							
2.14 Masonry							
2.14.1	BRICK/BLOCK WALLING						
2.14.1.1	Common brickwork in cement: lime mortar (1:1:6)						
2.14.1.1.1	Walls 102.5mm thick, brickwork; built against other work	196	m²				
2.14.1.1.2	Walls 215mm thick, brickwork	302	m²				

Figure A.6: Typical BQ format for a work package bill of quantities

Bill No. 8 EXTERNAL AND INTERNAL STRUCTURAL WALLS							
8.1 Masonry							
8.1.1	BRICK/BLOCK WALLING						
8.1.1.1	Common brickwork in cement: lime mortar (1:1:6)						
8.1.1.1.1	Walls 102.5mm thick, brickwork; built against other work	196	m²				
8.1.1.1.2	Walls 215mm thick, brickwork	302	m²				

A.5　Codifying items

As well as for the purpose of making it easy to search, it is essential that every item in the bill of quantities (BQ) can be referenced back to the cost plan.

A.6　Unit of measurement

The units of measurement for items are stipulated by the tabulated rules of measurement. For the purpose of clarity, the unit of measurement is to be entered against each item in the bill of quantities, irrespective of whether it is the same unit as the previous item.

A.7　Order of sizes

Sizes or dimensions in descriptions are to be in the order: length, width, height. Sometimes the width of a component (e.g. a base unit) is referred to as its 'depth'. If there is likely to be any doubt, for the purpose of clarity, the dimensions are to be stated.

For example:

Base unit: 1000mm long x 600mm wide x 900mm high; ...

A.8 Use of headings

Headings usually fall into one of four categories:

1 Elemental or work section headings

2 Element or sub-section headings

3 Headings that partly describe a group of items

4 Subdivisions required by *NRM 2: Detailed measurement for building works*.

A.9 Unit of billing

Other than enumerated and itemised items, the unit of measurement is the metre. The exception to this rule is steel bar reinforcement and structural steelwork, which are billed in tonnes to two decimal places.

A.10 Framing of descriptions

The bill of quantities is a legal document. Therefore, care should be taken when framing descriptions so that there is no doubt as to their meaning.

A.11 Totalling pages

There are a number of ways in which the quantity surveyor/cost manager might indicate how the cash totals on each page of the bill are to be dealt with. The preferred method is for the total to be carried over to be added to the next page and so on until the end of the bill or sub-section of the bill. Unless the bill section comprises only one page, the foot of the first and intermediate bill pages should be completed as follows:

Figure A.7: Examples of how to total pages

				Carried forward	£		

The top of the following bill page is completed as follows:

BILL 3: SUPERSTRUCTURE							
2.5 EXTERNAL WALLS				Brought forward	£		

To end each bill section, the section is completed as follows:

				TOTAL carried to main summary	£		

A.12 Price summary

Templates for the pricing summary for elemental bill of quantities (condensed and expanded versions) are included at Appendices D and E, respectively, of these rules. The structure of pricing summaries for other bill of quantities formats should follow the same principles.

Appendix B: Template for preliminaries (main contract) pricing schedule (condensed)

Cost Centre	Component	Time-Related Charges	Fixed Charges	Total Charges
		£ p	£ p	£ p
I	PRELIMINARIES			
1.1	EMPLOYER'S REQUIREMENTS			
1.1.1	Site accommodation			
1.1.2	Site records			
1.1.3	Completion and post-completion requirements			
1.2	MAIN CONTRACTOR'S COST ITEMS			
1.2.1	Management and staff			
1.2.2	Site establishment			
1.2.3	Temporary services			
1.2.5	Safety and environmental protection			
1.2.6	Control and protection			
1.2.7	Mechanical plant			
1.2.8	Temporary works			
1.2.9	Site records			
1.2.10	Completion and post-completion requirements			
1.2.11	Cleaning			
1.2.12	Fees and charges			
1.2.13	Site services			
1.2.14	Insurance, bonds, guarantees and warranties			
	Totals £			
	TOTAL CARRIED TO MAIN SUMMARY		£	

Note: Costs relating to main contractor's preliminaries items that are not specifically identified in the contractor's full and detailed breakdown shall be deemed to have no cost implications or have been included elsewhere within the contractor's rates and prices.

Appendix C: Template for preliminaries (main contract) pricing schedule (expanded)

Cost centre	Component	Time-related charges	Fixed charges	Total charges
		£ p	£ p	£ p
1.1	EMPLOYER'S REQUIREMENTS			
1.1.1	Site accommodation			
1.1.1.1	Site accommodation			
1.1.1.2	Furniture and equipment			
1.1.1.3	Telecommunications and IT systems			
1.1.2	Site records			
1.1.2.1	Site records			
1.1.3	Completion and post-completion requirements			
1.1.3.1	Handover requirements			
1.1.3.2	Operation and maintenance services			
1.2	MAIN CONTRACTOR'S COST ITEMS			
1.2.1	Management and staff			
1.2.1.1	Project specific management and staff			
1.2.1.2	Visiting management and staff			
1.2.1.3	Extraordinary support costs			
1.2.1.4	Staff travel			
1.2.2	Site establishment			
1.2.2.1	Site accommodation			
1.2.2.2	Temporary works in connection with site establishment			
1.2.2.3	Furniture and equipment			
1.2.2.4	IT systems			
1.2.2.5	Consumables and services			
1.2.2.6	Brought-in services			
1.2.2.7	Sundries			
1.2.3	Temporary services			
1.2.3.1	Temporary water supply			
1.2.3.2	Temporary gas supply			
1.2.3.3	Temporary electricity supply			
1.2.3.4	Temporary telecommunication systems			
1.2.3.5	Temporary drainage			
1.2.4	Security			
1.2.4.1	Security staff			
1.2.4.2	Security equipment			
1.2.4.3	Hoardings, fences and gates			
1.2.5	Safety and environmental protection			
1.2.5.1	Safety programme			
1.2.5.2	Barriers and safety scaffolding			
1.2.5.3	Environmental protection measures			

1.2.6	Control and protection			
1.2.6.1	Survey, inspections and monitoring			
1.2.6.2	Setting out			
1.2.6.3	Protection of works			
1.2.6.4	Samples			
1.2.6.5	Environmental control of building			
1.2.7	Mechanical plant			
1.2.7.1	Generally			
1.2.7.2	Tower cranes			
1.2.7.3	Mobile cranes			
1.2.7.4	Hoists			
1.2.7.5	Access plant			
1.2.7.6	Concrete plant			
1.2.7.7	Other plant			
1.2.8	Temporary works			
1.2.8.1	Access scaffolding			
1.2.8.2	Temporary works			
1.2.9	Site records			
1.2.9.1	Site records			
1.2.10	Completion and post-completion requirements			
1.2.10.1	Testing and commissioning plan			
1.2.10.2	Handover			
1.2.10.3	Post-completion services			
1.2.11	Cleaning			
1.2.11.1	Site tidy			
1.2.11.2	Maintenance of roads, paths and pavings			
1.2.11.3	Building clean			
1.2.12	Fees and charges			
1.2.12.1	Fees			
1.2.12.2	Charges			
1.2.13	Site services			
1.2.13.1	Temporary works			
1.2.13.2	Multi-service gang			
1.2.14	Insurance, bonds, guarantees and warranties			
1.2.14.1	Works insurance			
1.2.14.2	Public liability insurance			
1.2.14.3	Employer's (main contractor's) liability insurance			
1.2.14.4	Other insurances			
1.2.14.5	Bonds			
1.2.14.6	Guarantees			
1.2.14.7	Warranties			
	Totals £			
	TOTAL CARRIED TO MAIN SUMMARY		£	

Note: Costs relating to preliminaries items that are not specifically identified in the contractor's full and detailed breakdown shall be deemed to have no cost implications or have been included elsewhere within the contractor's rates and prices.

Appendix D: Template for pricing summary for elemental bill of quantities (condensed)

Cost centre	Element	£/p	£/p
0.0	Facilitating works		£0.00
1.0	Substructure		£0.00
2.0	Superstructure		£0.00
3.0	Internal finishes		£0.00
4.0	Fittings, furnishings and equipment		£0.00
5.0	Services		£0.00
6.0	Prefabricated buildings and building units		£0.00
7.0	Work to existing building		£0.00
8.0	External works		£0.00
	TOTAL (Building works, including M&E engineering services)		£0.00
9.0	Main contractor's preliminaries		£0.00
	Sub-total		£0.00
10.0	Provisional sums:		£0.00
10.1	Defined provisional sums	£0.00	
10.2	Undefined provisional sums	£0.00	
10.3	Works to be carried out by statutory undertakers	£0.00	
	Sub-total		£0.00
11.0	Risks		£0.00
11.1	Sub-total		£0.00
12.0	Main contractor's overheads and profit (insert required % adjustment)	0.00%	£0.00
	Sub-total		£0.00
13.0	Credit (for retained arisings)		£(0.00)
	Sub-total		£0.00
14.0	Main contractor's fixed price adjustment (insert required % adjustment)	0.00%	£0.00
	Sub-total		£0.00
15.0	Director's adjustment (insert required adjustment (+/-))		£0.00 or £(0.00)
	Sub-total		£0.00
16.0	Dayworks (Provisional)		£0.00
	TOTAL TENDER PRICE, exclusive of VAT (Carried to Form of Tender)		£0.00

Appendix E: Template for pricing summary for elemental bill of quantities (expanded)

Cost centre	Element	£/p	£/p
0.0	Facilitating works		£0.00
0.1	Toxic/hazardous/contaminated material treatment	£0.00	
0.2	Major demolition works	£0.00	
0.3	Specialist ground works	£0.00	
0.4	Temporary diversion works	£0.00	
0.5	Extraordinary site investigation works	£0.00	
1.0	Substructure		£0.00
1.1	Substructure	£0.00	
2.0	Superstructure		£0.00
2.1	Frame	£0.00	
2.2	Upper floors	£0.00	
2.3	Roof	£0.00	
2.4	Stairs and ramps	£0.00	
2.5	External walls	£0.00	
2.6	Windows and external doors	£0.00	
2.7	Internal walls and partitions	£0.00	
2.8	Internal doors	£0.00	
3.0	Internal finishes		£0.00
3.1	Wall finishes	£0.00	
3.2	Floor finishes	£0.00	
3.3	Ceiling finishes	£0.00	
4.0	Fittings, furnishings and equipment		£0.00
4.1	Fittings, furnishings and equipment	£0.00	
5.0	Services		£0.00
5.1	Sanitary installations	£0.00	
5.2	Services equipment	£0.00	
5.3	Disposal installations	£0.00	
5.4	Water installations	£0.00	
5.5	Heat source	£0.00	
5.6	Space heating and air conditioning	£0.00	
5.7	Ventilation	£0.00	
5.8	Electrical installations	£0.00	
5.9	Fuel installations/systems	£0.00	
5.10	Lift and conveyor installations/systems	£0.00	
5.11	Fire and lightning protection	£0.00	
5.12	Communication, security and control systems	£0.00	
5.13	Special installations/systems	£0.00	
5.14	Builder's work in connection with services	£0.00	
6.0	Complete buildings		£0.00

6.1	Pre-fabricated buildings		£0.00	
7.0	**Work to existing building**			£0.00
7.1	Minor demolition works and alteration works		£0.00	
7.2	Repairs to existing services		£0.00	
7.3	Damp proof courses/fungus and beetle eradication		£0.00	
7.4	Facade retention		£0.00	
7.5	Cleaning existing surfaces		£0.00	
7.6	Renovation works		£0.00	
8.0	**External works**			£0.00
8.1	Site preparation works		£0.00	
8.2	Roads, paths and pavings		£0.00	
8.3	Soft landscaping, planting and irrigation systems		£0.00	
8.4	Fencing, railings and walls		£0.00	
8.5	Site/street furniture and equipment		£0.00	
8.6	External drainage		£0.00	
8.7	External services		£0.00	
8.8	Minor building works and ancillary buildings		£0.00	
	TOTAL (Building works, including M&E engineering services)			£0.00
9.0	**Main contractor's preliminaries**			£0.00
	Sub-total			£0.00
10.0	**Risks**			£0.00
	Sub-total			£0.00
11.0	**Provisional sums:**			£0.00
11.1	Defined provisional sums		£0.00	
11.2	Undefined provisional sums		£0.00	
11.3	Works to be carried out by statutory undertakers		£0.00	
	Sub-total			£0.00
12.0	**Main contractor's overheads and profit** (insert required % adjustment)	0.00%		£0.00
	Sub-total			£0.00
13.0	**Credit** (for retained arisings)			£(0.00)
	Sub-total			£0.00
14.0	**Main contractor's fixed price adjustment** (insert required % adjustment)	0.00%		£0.00
	Sub-total			£0.00
15.0	**Director's adjustment** (insert required adjustment (+/-))			£0.00 or £(0.00)
	Sub-total			£0.00
16.0	**Dayworks** (Provisional)			£0.00
	TOTAL TENDER PRICE, exclusive of VAT (Carried to Form of Tender)			£0.00

Appendix F: Templates for provisional sums, risks and credits

Schedule of provisional sums

Cost centre	Provisional sum	£/p
	DEFINED PROVISIONAL SUMS	
	UNDEFINED PROVISIONAL SUMS	
	TOTAL PROVISIONAL SUMS, exclusive of VAT (Carried to Main Summary)	

Schedule of construction risks

Cost centre	Risk description	£/p
R001		
R002		
R003		
R004		
R005		
R006		
	TOTAL RISK ALLOWANCE, exclusive of VAT (Carried to Main Summary)	

Credits

Cost centre	Description	£/p
C001		
C002		
C003		
C004		
C005		
	TOTAL CREDITS, exclusive of VAT (Carried to Main Summary)	

Appendix G: Example of a work package breakdown structure

Figure 2.5: Typical example of suffix codes used for codifying work packages when a work package BQ breakdown structure is used

Serial no.	Work package title/content	Suffix
1.	Main contractor's preliminaries	/01.1
2.	Intrusive investigations: Asbestos and other hazardous materials Geotechnical and environmental investigations Attendance on archaeological investigations Work package contractor's preliminaries	/02 /01.2
3.	Demolition works: Asbestos and other hazardous materials removal/treatment works Soft strip of building components and sub-components Soft strip of mechanical and electrical engineering services. Demolition. Work package contractor's preliminaries	/03 /01.2
4.	Groundworks: Contaminated ground material removal; Preparatory earthworks; Excavation and earthworks, including basement excavation, earthwork support and disposal; Temporary works – propping of existing basement retaining walls; Below ground drainage; Ground beams; Pile caps; Temporary works – piling mats/platforms; Ground bearing base slab construction, including waterproofing; and Basement retaining wall structures, including waterproofing. Work package contractor's preliminaries	/04 /01.2
5.	Piling: Piling works Work package contractor's preliminaries	/05 /01.2
6.	Concrete works: Frame Upper floors, including roof structure Core and shear walls Staircases Work package contractor's preliminaries	/06 /01.2
7.	Roof coverings and roof drainage: Roof cladding/coverings; Flashings; and Roof drainage. Work package contractor's preliminaries	/07 /01.2

Serial no.	Work package title/content	Suffix
8.	**External and internal structural walls:** Structural steelwork Masonry (brickwork and blockwork) Roof systems and rainwater goods Cladding Curtain walling Carpentry General joinery Bespoke joinery Windows and external doors Dry linings and partitions Architectural metal work Work package contractor's preliminaries	/08 /01.2
9.	Cladding: Cladding systems, including integral windows and external doors. Work package contractor's preliminaries	/09 /01.2
10.	**Windows and external doors:** (Non-integral to cladding system) Windows; Louvers; External doors; and Shop fronts. Work package contractor's preliminaries	/10 /01.2
11.	**Mastic:** Mastic to windows, louvers and external door frames; and Mastic to wet areas. Work package contractor's preliminaries	/11 /01.2
12.	**Non-structural walls and partitions:** Tiling (floor and wall) Internal stone finishes Painting and decorating Soft floor coverings Suspended ceilings Work package contractor's preliminaries	/12 /01.2
13.	**Joinery:** Reception desk Internal door sets; Screens; Toilet cubicles; Timber wall linings to toilet cubicles; Skirtings; and All other second fix joinery items. Work package contractor's preliminaries	/13 /01.2
14.	**Suspended ceilings:** Suspended ceilings Work package contractor's preliminaries	/14 /01.2
15.	**Architectural metal work:** All architectural metal work items Work package contractor's preliminaries	/15 /01.2
16.	**Tiling:** Internal stone finishes Wall tiling; and Floor tiling Work package contractor's preliminaries	/16 /01.2

Serial no.	Work package title/content	Suffix
17.	**Painting and decorating:** Painting and decorating Work package contractor's preliminaries	/17 /01.2
18.	**Floor coverings:** Carpet; and Vinyl tiles. Work package contractor's preliminaries	/18 /01.2
19.	**Fittings, furnishings and equipment:** Cupboards and shelves to storerooms Loose fittings, furnishings and equipment; and Signage Work package contractor's preliminaries	/19 /01.2
20.	**Combined Mechanical and Electrical Engineering Services:** Sanitary appliances, including kitchette sinks; Mechanical engineering services installations; Electrical engineering services installations; Public health engineering services installations (above ground); and Lifts (by named subcontractor). Work package contractor's preliminaries	/20 /01.2
21.	**Lifts and escalators:** Passenger lifts; Fire fighting lift; Platforms; and Escalators	/21 /01.2
22.	**Facade access equipment:** Building maintenance units (BMUs), including proprietary storage units. Work package contractor's preliminaries	/22 /01.2
23.	**External works and drainage:** External drainage Soft landscape works Hard landscape works Work package contractor's preliminaries.	/231 /01.2

Bibliography

Building Costs Information Service: *Standard Form of Cost analysis* (4th edition), BCIS, 2012

Published by the Office of Government Commerce, ITIL, 2007:

- OGC Gateway™ Process *Review 0: Strategic assessment*
- OGC Gateway™ Process *Review 1: Business justification*
- OGC Gateway™ Process *Review 2: Delivery strategy*
- OGC Gateway™ Process *Review 3: Investment decision*
- OGC Gateway™ Process *Review 4: Readiness for service*
- OGC Gateway™ Process *Review 5: Operational review and benefits realisation*

Royal Institute of British Architects, *RIBA Outline Plan of Work 2007* (amended November 2008), RIBA, 2008

Royal Institution of Chartered Surveyors, *Cost Management in Engineering Construction Projects*, RICS, Surveyors Holdings Limited, London, 1992

Index

A

E

T

W

easy to read
goodnight stories

Brown Watson
ENGLAND

Contents

First Published 1994 by Brown Watson
The Old Mill, 76 Fleckney Road,
Kibworth Beauchamp, Leics LE8 0HG
© 1994 Brown Watson, England
Reprinted 2000, 2001, 2003, 2005, 2006
ISBN: 0-7097-1339-8

Printed in China

Pinocchio

Geppetto was a poor toymaker whose dearest wish was to have a son. One day, as he sat making a wooden puppet, it seemed to look at him and to smile. "How I wish I could look on the face of my son," he said. "I would call him Pinocchio."

Geppetto did not know it, but the Blue Fairy had heard what he said. "He deserves to have his wish granted," she thought. "Pinocchio shall be a son to Geppetto."

And, as Pinocchio's eyes opened wide, so there came a chirruping noise from the fireplace. "Meet Jiminy Cricket!" said the fairy. "He is your conscience to tell you right from wrong, Pinocchio."

Geppetto was overjoyed to have a son at last! "You must go to school, Pinocchio," he said.

"That's right!" nodded Jiminy Cricket. "Otherwise, you'll turn into a donkey."

Geppetto even sold his only jacket so that he could afford to buy Pinocchio the spelling book he needed to take to school.

"Goodbye, Father!" he called. "I shall make you proud of me."

But that was before Pinocchio knew Fire-Eater's Puppet Theatre was in town! Taking no notice of Jiminy Cricket, he sold his book to buy a ticket – and soon he was on stage, singing and dancing.

Fire-Eater wanted Pinocchio to stay. But when the time came to move far away from home and Geppetto, he was afraid. Being a wonderful singing, dancing puppet didn't seem so clever, after all. . .

"Geppetto sold his jacket to send me to school," he sobbed to Jiminy Cricket. "He'll wonder where I am!"

Luckily for him, Fire-Eater knew Geppetto and he gave Pinocchio five pieces of gold to take home.

"I can buy Geppetto a new jacket," cried Pinocchio. "Five gold pieces!"

"Is that all?" scoffed a cat.

"Bury them in our magic field," said the fox with him. "You'll have a tree of gold next day!"

"No, Pinocchio!" said Jiminy Cricket. "That's Geppetto's money!"

How Pinocchio wished he had listened to Jiminy when he discovered that the crafty cat and the sly fox had dug up the gold he had buried!

The fairy heard Pinocchio crying and asked him what was wrong.

"I dropped the gold I was taking home to Geppetto," he sobbed. "Now I can't find it!" And as he spoke, something very strange happened. . .

Pinocchio's nose began to grow!
"Where do you think you lost the money?" asked the fairy.
"On the way to school," he cried.
His nose grew even longer!
"It fell out of my pocket."

By now, his nose was so long, he
could hardly see the end of it!

"Well, Pinocchio," laughed the
Blue Fairy, "now you know how one
small lie can grow into a big lie —
just like your nose!"

At once, Pinocchio promised not to tell any more lies, sobbing so hard that the fairy took pity on him. "If you had listened to Jiminy Cricket," she said, "none of this would have happened !"

Pinocchio knew this was true, and, full of good intentions, he set off home. He had only gone a little way when a carriage full of children came along, pulled by some strange-looking donkeys!

"Come to the Land of Toys and play all the year round!" they cried.

"Don't listen to them," warned Jiminy Cricket. But Pinocchio was already jumping up, determined not to miss any of the fun.

He thought the Land of Toys was wonderful! No books, no lessons – just as much play as anyone wanted!

But after a while, he noticed his ears felt rather heavy – heavy and long, thick and furry . . .

"I said that you'd turn into a donkey if you didn't go to school," scolded Jiminy Cricket. "What will you do, now?"

"Geppetto!" cried Pinocchio. "I want to go home to Geppetto!"

Pinocchio was afraid everyone would laugh at his donkey ears. But the people were too upset even to notice. "Geppetto went to sea, looking for you," they said. "We think he was swallowed by a whale!"

"Poor Father," cried Pinocchio. "I must find him!" He made his way to the place where Geppetto was last seen and jumped into the inky blackness of the sea, gusts of wind hitting him in the face.

Suddenly, he saw a light ahead. He swam towards it and found himself crawling, then walking into a sort of underground cavern. "Pinocchio!" cried a voice. "Pinocchio, my dear, brave son!"

Pinocchio had never been so glad to see anyone – even if he had swum inside a whale by mistake!

"We'll get through the whale's mouth, then make for the shore," he told Geppetto. "Just follow me!"

When the whale opened its mouth, they quickly swam out! But Pinocchio was soon very tired, swimming for Geppetto as well as himself. By the time Jiminy Cricket had guided him to dry land, he could hardly move.

The Blue Fairy was waiting when Geppetto carried him to dry land.

"Well done, Pinocchio," she said. "You have shown that you are a brave and loving son. You shall have your reward!"

And instead of a little wooden puppet, Pinocchio became a real boy with a beaming smile for Geppetto – and a conscience of his own to tell him right from wrong. How happy Jiminy Cricket was for both his friends!

Alice in Wonderland

Alice was tired of sitting on the bank. The sun made her feel sleepy. Suddenly a white rabbit ran past saying, "Oh, dear! Oh, dear! I shall be too late!" and taking a watch from his waistcoat pocket!

Alice had never seen a white rabbit with a waistcoat, or a pocket watch, so she followed him to see where he was going. And when the White Rabbit went down a rabbit hole – down went Alice after it.

Suddenly, Alice felt herself falling, down, down, down . . . until, thump! She landed on a heap of dry leaves. "Oh, my ears and whiskers!" she heard the White Rabbit saying. "How late it is getting!"

All at once, he vanished from sight – leaving Alice in a long, low hall with locked doors all around. Alice came across a little table on which there was a key. But, which door did it fit?

Then she came across a tiny door behind a curtain. She turned the key in the lock, and the door opened. Kneeling down, she could see a beautiful garden, with bright flower beds and water fountains.

"Now," thought Alice, "how do I get out?" On the table where the key had been there was a bottle with "DRINK ME!" written on a label around the neck – it was the most delicious drink Alice had ever tasted!

And with every drop, Alice became smaller and smaller – until she was just the right size to get through the little door and into the beautiful garden. If only she had not left the key on the table . . .

Then, she found a box with a cake inside. In currants were the words, "EAT ME." So, Alice did, growing taller and taller – until she was much too big to go through the door, even though she had the key!

Alice was so sad, she began to cry. Before long, there was the patter of feet and in came the White Rabbit. The sight of such a big, tall Alice frightened him so much, he dropped the fan and gloves he was carrying.

Alice picked up the fan, quickly dropping it when she found herself shrinking again! Her foot slipped, and – splash! She was up to her chin in the pool of tears she had wept when she had been so tall.

In the pool of tears were also a duck, a dodo – and a mouse who told a story to get everyone dry!

"Mary Ann!" called a voice. "Fetch me my gloves this minute!" It was the White Rabbit.

This time, Alice followed him to a little house with "W. RABBIT" on the door. Coming out the other side, she saw a huge mushroom, about as tall as she was now, on which a caterpillar sat, smoking a pipe!

The caterpillar said that eating
one side of the mushroom would
make her taller, the other side
smaller. So Alice took a piece of
each. A Cheshire Cat grinned at her
from a tree.

"Please," said Alice to the cat, "which way should I go?"

"That way," said the cat waving his right paw, "lives the Mad Hatter, and that way," waving his other paw, "lives the March Hare!"

The March Hare's house had a roof thatched with fur and chimneys shaped like long ears! He and the Mad Hatter sat outside at a table, resting their elbows on a very sleepy dormouse.

"Tell us a story!" ordered the March Hare.

"Well," said Alice, taken by surprise, "I don't think . . ."

"Don't think?" he echoed. "Then you shouldn't talk!"

That was enough to make Alice march away from the table. Quite by chance she saw one of the trees had a door set in it. When Alice opened it, she was in the hall again.

Nibbling one piece of the caterpillar's mushroom, then the other, Alice made herself the right size to get the key, then go through the little door and out into the beautiful garden, at last!

In the garden, two gardeners were painting white roses red! "We planted a white rose by mistake!" explained one, "and if the queen finds out . . . . Oh no! Here she comes  now!"

The Queen of Hearts stopped when she saw Alice. "What is your name, child?" she asked.

"My name," said Alice, "is Alice."

"Can you play croquet?"

"Oh, yes!" cried Alice.

Alice had never played croquet using flamingoes to hit curled-up hedgehogs! Before long the game was a real mess, with the queen yelling "Off with his head!" or "Off with her head!" every other minute.

Suddenly, Alice heard a cry.
"The trial is beginning!"
"What trial?" asked Alice – but
everyone was already running
ahead, carrying her along with
them into a big court room.

"The charge is," said the White
Rabbit, "that the Knave of Hearts
stole some tarts!"

"Call the first witness!" said the
king. And to her great surprise, the
White Rabbit called, "Alice!"

"What do you know of this?" asked the king.

"Nothing whatsoever," said Alice.

"Off with her head!" shouted the Queen of Hearts, red in the face.

"You?" went on Alice. "You're only a pack of cards!"

At once, the cards rose up and came flying down on her! Alice gave a little scream – and found herself on the bank with her sister.

Her adventures in Wonderland had only been a wonderful dream!

The Ugly Duckling

Mother Duck had found the perfect place on the farm to build her nest. It was next to a little stream. The other ducks quacked and splashed about in the water, as she sat waiting for her eggs to hatch.

Then, at last, came the great day when the shells burst open, one after the other!

"Cheep-Cheep!" piped the tiny, yellow ducklings as they waddled around. "How wide the world is!"

Mother Duck fussed round them proudly. She was so busy trying to keep all her ducklings together, that, at first, she did not notice there was one egg, bigger than any of the others, still in the nest.

"That's a turkey's egg!" quacked an old duck when she saw it. "And turkeys never learn to swim, my dear, not like our little ones. Nasty birds, they are, too! Take my advice and leave it alone."

But Mother Duck said she would sit on the egg a little longer until it hatched. And, instead of a pretty, yellow duckling, out came a fat, ugly chick with horrible dark grey feathers!

"Was this really a turkey chick?" wondered Mother Duck, leading the way down to the stream. How glad she was to see the ugly, little bird swimming along behind the others.

"He's not a turkey," she thought, "just an Ugly Duckling."

The Ugly Duckling soon began to grow, and as he grew, the uglier he became. The other ducklings wouldn't even talk to him.

The hens in the farmyard pecked at him whenever he came near. Worst of all was the turkey cock, who came at the Ugly Duckling making loud gobbling noises, until it was red in the face.

Even the little girl who fed the farmyard birds aimed kicks at him. Unhappy and frightened, he flew off, some smaller birds getting out of his way. "That's because I'm so ugly," he thought.

The Ugly Duckling flew on until he came to a marsh where some wild ducks lived.

"My," said one, "you're so ugly!"

The Ugly Duckling just fluffed up his feathers and fell asleep.

Next day, the air was shattered by hunters shooting at the wild ducks. The Ugly Duckling thought he would die when one of the dogs found him. Then – splash – the dog turned and went.

"I'm so ugly!" thought the Ugly Duckling. "Even the dog does not bite me." And he went on his way, until he came to a hut where an old woman lived with her cat and a hen. He crept inside.

The woman thought the Ugly Duckling was a lady duck to lay eggs for her. But as he grew fatter and uglier and no eggs came, she got angry. The hen and cat hated him because he could swim.

All through the summer the Ugly Duckling was all alone, eating whatever he could find. Then came the autumn, when the leaves blew down from the trees and the clouds hung low in the sky.

Then, at sunset one day, the Ugly Duckling saw the most beautiful white birds flying across the lake. He watched them until they were out of sight, wishing with all his heart that he could be with them.

The winter snow reminded him of those beautiful white birds. The river froze, almost freezing the Ugly Duckling with it, until a kind man broke the ice and took him home.

His children wanted to play, but the Ugly Duckling thought they would hurt him. They scared him so much that he splashed into a pail of milk, and then into a barrel of oatmeal!

The man's children laughed and laughed, but their mother was furious. The Ugly Duckling only just missed being hit by the fire tongs, as he ran out into the bitter winter weather.

Now came the worst part of the Ugly Duckling's whole life. Often he felt he would die from hunger and cold, longing for some shelter. He could hardly believe it when the sun shone again and the birds sang.

Hearing the birds, the Ugly Duckling flapped his own wings, surprised to find how big and strong they had become. The sun warmed his back as he flew, making him feel

happier than he had been for a long, long time.

On and on flew the Ugly Duckling until he saw a garden, the scent of flowers wafting up towards him. Suddenly, three beautiful white swans flew out from the thicket, gliding into the water.

These were the birds he had seen in the autumn, the ones he loved – although he did not know why. "What if they hurt me?" he thought. "Better to die here than to be beaten and punished because I'm so ugly . . ."

Slowly, the swans turned and came towards him, looking so solemn that the Ugly Duckling bowed his head. He saw his reflection in the water – not the reflection of an Ugly Duckling but of a beautiful white swan.

The Ugly Duckling thought he was dreaming! Could he really be a beautiful swan?

"There's a new swan! Isn't he lovely?" said some children as they stood by  the lake.

The handsome young swan lifted his head, looking all around him.

"This cannot be a dream," he thought. "I could never, ever have dreamed of being so happy when I was the Ugly Duckling!"

Goldilocks
and the Three Bears

There was once a girl whose hair was so fair and curly, that everyone called her Goldilocks.

Goldilocks and her family lived near a forest, and she liked nothing better than going for long walks on her own.

Goldilocks thought she must know every inch of that forest until, one morning, after she had set off a little earlier than usual, she saw something which gave her quite a surprise. . .

It was a little cottage she had never seen before, with lace curtains at the windows and smoke coming out of the chimney.

"Who can live here?" wondered Goldilocks, going up to the door.

She knocked at the door and waited. There was no answer. She knocked again. Still, no answer.

"Anyone at home?" she called, and knocked again, a little harder this time. The door creaked open.

Goldilocks stepped inside and looked all round such a cosy, little room. A fire burned cheerfully, and on the hob were three bowls of porridge – a big bowl, a smaller bowl, and a tiny, little bowl. . .

"I wonder who lives here?" thought Goldilocks again, never guessing it was the home of three bears – Daddy Bear, Mummy Bear and Baby Bear. She only knew how good that porridge looked on a fresh, spring morning.

First she tasted Daddy Bear's porridge. That was too hot. Then, she tried Mummy Bear's porridge. That was too cold. But when she tasted Baby Bear's porridge, it was so good that Goldilocks soon ate it all up!

After eating all that porridge, Goldilocks wanted to sit down. So she tried Daddy Bear's chair. That was too hard. Then she tried Mummy Bear's chair, but that was too soft. Then, she tried Baby Bear's chair. . .

And that was just right! In fact, Goldilocks had never sat in such a comfortable chair! She wriggled and squirmed so much, that, in the end, the chair broke, and Goldilocks fell to the floor!

"Ooh!" she groaned. "I think I'd better go and lie down." So, she went upstairs.

And in the bedroom were three beds – Daddy Bear's bed, Mummy Bear's bed, and Baby Bear's bed. . .

First, she tried Daddy Bear's bed. But that was too hard. Then she tried Mummy Bear's bed. That was too soft. But Baby Bear's bed was so warm and so cosy that Goldilocks snuggled down and was soon fast asleep!

By this time, Daddy Bear, Mummy Bear and Baby Bear were coming back from their walk. They had only gone to the end of the forest path and back – "Just to let the porridge cool down," Mummy Bear had said.

"Who's been eating my porridge?" growled Daddy Bear.

"Who's been eating my porridge?" said Mummy Bear.

"Who's been eating my porridge?" cried Baby Bear. "There's none left!"

"And who's been sitting in my chair?" roared Daddy Bear.

"Who's been sitting in my chair?" cried Mummy Bear.

"Who's been sitting in my chair?" wailed Baby Bear. "It's all broken!"

They went upstairs. "Who's been sleeping in my bed?" said Daddy Bear.

"Who's been sleeping in my bed?" squealed Mummy Bear.

"Who's been sleeping in my bed?" said Baby Bear, with a loud sob.

His cries woke Goldilocks and she sat straight up in bed. She could not believe her eyes when she saw three furry faces looking at her! "B-bears!" she blurted out, very frightened. "Th-three b-bears!"

Had Goldilocks known it, Daddy, Mummy and Baby Bear were gentle, kind bears. When they saw it was only a little girl who had been in their cottage, they were not nearly so angry as they might have been.

But Goldilocks only knew that she had to leave their cottage just as soon as she could. So she let out a scream, the loudest, longest scream she had ever screamed, making the Three Bears jump back at once!

This was Goldilocks' chance! She flung back the bedclothes and rushed out of the door and down the stairs, away back into the forest before the Three Bears knew what was happening!

On and on she ran through the forest until she felt she could run no more. It seemed a long, long time before she reached the path which led to her own home.

And there was her mother, waiting

anxiously at the gate. Goldilocks was so glad to see her.

"Oh, where have you been, Goldilocks?" she cried. "Daddy was just going out to look in the forest for you!"

And so began the story of Goldilocks and the Three Bears. Her mother could hardly believe it!

"You naughty girl!" she scolded. "Haven't I always told you never to go inside strange places?"

"Goldilocks," said her daddy, "are you sure this tale about The Three Bears isn't an excuse because you do not know the forest as well as you thought?"

"No, Daddy," cried Goldilocks.

"Here," she went on, taking his hand, "I'll take you to their cottage, myself. Then you'll see."

And she quickly led the way back into the forest without stopping once, seeming sure of every step.

That was the first of many times Goldilocks went back to the forest. But, no matter how hard she searched, she did not find that little cottage, nor The Three Bears – Daddy Bear, Mummy Bear and Baby Bear.